A Gender Perspective of Municipal Solid Waste Generation & Management in the City of Bamenda, Cameroon

Akum Hedwig Kien

T0282323

Langaa Research & Publishing CIG
Mankon, Bamenda

Publisher:
Langaa RPCIG
Langaa Research & Publishing Common Initiative Group
P.O. Box 902 Mankon
Bamenda
North West Region
Cameroon
Langaagrp@gmail.com
www.langaa-rpcig.net

Distributed in and outside N. America by African Books Collective
orders@africanbookscollective.com
www.africanbookscollective.com

ISBN-10: 9956-550-63-9

ISBN-13: 978-9956-550-63-0

Dedication

To my dear husband and friend

Gideon Akumah Ngwa,

and loving Children.

Preface

I was pushed into researching on waste management issues by my experiences living in a typical growing and expanding urban area in a developing economy. In the town of Bamenda where I grew up, it is was common place to find pedestrians, cars and rubbish tips piled up with unsorted waste from households and markets competing for space on narrow roads. Insects and rodents foraged at the waste sites and stench ensued from them. Gutters and water courses were blocked with waste. Observation indicated the effective presence of women in household and market activities that had bearings with waste generation. Then it struck me that there could be a gender angle to the generation and management of municipal waste that was not exploited. The gap was made even more apparent by the fact that environmental and gender concerns take central stage in contemporary discourses, yet urban waste management has remained a challenge to the municipal authorities of Cameroonian cities and towns. The question then is whether the problem can be thoroughly researched into towards a coherent framework for municipal solid waste management.

Unprecedented urbanisation and life style changes that drift towards a use and throw-away society continue to be a major contributing factor for waste increase. Waste characterisation including biodegradable and non-biodegradable with plastics and electronic waste streams further compound the challenges of waste management efforts. Unfortunately, these upsurges in waste production are often not accompanied by a commensurate improvement in management efforts. Hence, waste dumping with consequent environmental implications is common place in most cities.

Municipal Waste management problems are often associated with inadequate finance and technology, poor planning, lack of expertise and other resources. However, we posit that a concerted social enculturation through sensitisation and awareness campaigns can assuage the problem. Waste management strategies have focused on waste collection, transportation and disposal by the council or organisations while ignoring waste generation. By so doing, the pools and actors of waste generation and primary level management are overlooked. The activities, the gender dynamics and politics at the pools of waste generation, particularly the households and markets largely influence the success or failures of waste management strategies and policies.

This work brings out the gender dimension of municipal waste generation and management. It makes visible the role of women, men and children in the municipal solid waste management problem. It is hoped that the findings revealed and proposals made from the study will be employed by municipal authorities in Bamenda and beyond to enhance waste management efforts.

This book has been developed from a PhD thesis that I completed for the gender and development programme in the Department of Women and Gender Studies of the University of Buea, Cameroon. The major findings suggest that municipal solid waste generation activities in households and market places are gendered with women playing leading roles. That waste activities at the primary level of management are championed by women and children. Though municipal waste generation activities are gendered, council waste management strategies are not gendered. That there is a large gap between environmental related awareness and environmental consciousness in people's behaviour as concerns waste management issues. Mainstreaming gender in waste management policies and strategies alongside proper planning, adequate finances and technology will go a long way to improve on waste management and urban sanitation.

To make the findings and proposals of the study to reach a wider audience, data from the thesis, and some parts of this book, have already been published in article form in peer reviewed journals. However, the core of the work in the original thesis in the form defended before a constituted academic panel of the University is presented in this book. Anyone who wants to have a feel and understanding of the gender roles played by the different actors in municipal solid waste generation and management will find this book very useful.

Akum Hedwig Kien
Buea
October 2018

Table of Contents

List of Illustrations

Tables

Figures

Plates

Maps

Acknowledgements

I am grateful to Professor William Markham and Lotsmart Fonjong, who relentlessly supervised this work and read over the multiple versions of the draft copies, their invaluable critique, insightful, inspirational and encouraging comments have positively impacted this work.

I am thankful to the Department of Women and Gender Studies, staff and colleagues for collaboration and encouragement. Special thanks to Professor J.B. Endeley, Dr Abonge Christiana (acting Head of Department), Justine Ayuk and Dorothy Oben for their time and selfless contributions.

Gratitude goes to informants in the field for very valuable contributions. I will forever be grateful to the Chief of Service for Sanitation and Hygiene at the BCC, Ms Julie Chambi. Due appreciation is given to the research assistants for their time and diligence in collecting data. Thanks to the statistician, Mr Nana who was very instrumental in formatting research instruments and for patience in analysing the data.

For editing, and other insightful opinions my appreciation goes to Mr Ngi Christopher, Gideon Akumah Ngwa (Ass. Prof), Dr Neba Ayu'nwi and Mrs FuhSuh Aurelia.

Hearty appreciation to my family members particularly, my husband, the children (Ade-Ngwa, Ndeh-Ngwa, Kien-Ngwa, Ngang and Desmond), Derick, Goddy, Pascaline, Barbara, Vicky and Akum, for their support and even for distractions. To my mother, Mama Dorothy Akoba, Mr Akum Michael, Nghia Bern and Ivo, I say thank you immensely for always being there for me. To all my relations, friends and everyone who accompanied me through this academic journey, I will be forever grateful.

Abstract

The management of urban waste constitutes one of the major environmental challenges facing African cities in general and Cameroon in particular. Unprecedented population growth and changes in consumption patterns and lifestyles have led to increased waste generation. Municipal solid waste management efforts take up as much as 50 percent of available budgets of some municipalities but lag behind the rate of waste generation with attendant environmental and public health risks. Undesirable waste practices of urban citizens have negatively compounded the plight of the urban poor (especially women) as portrayed by the effects of poor garbage disposal and, air and water pollution in overcrowded, unplanned urban cities. The management of urban waste by the actors involved must begin with a mastery of waste generation and handling practices; particularly domestic waste which accounts for more than 70% of all waste in most African cities. Global development discourses consider gender and environment issues as interrelated and crucial to the success of sustainable development efforts, including waste generation and management. While past research on waste problems in Bamenda has focused on population growth, limited funding, and inadequate technology and expertise as the main causes of poor waste management, this thesis argues that the omission of a gender perspective in municipal waste policies and strategies is another major barrier to efficient waste management. There has been little research into the relationship between household gender politics, the gender division of labour and gender roles, and the inadequacies of municipal solid waste management. This thesis therefore investigates the influence of gender on waste generation and management and the extent to which the council considers this relationship in its waste management policies and strategies. The study focused on two main sources of municipal solid waste, households and markets. The population covered by the study consisted of 47,585 households and nine markets in the urban area controlled by the Bamenda City Council. At a confidence interval of

95 percent, with an expected prevalence of 80 percent a sample size of 367 households was estimated. To this number, results from 339 households were effectively retained for the study giving a 92 percent response rate. A cross-sectional survey based on a multi-stage sampling technique ensured proper geographical coverage and representation of the households. Markets located in the three subdivisions of Bamenda I, Bamenda II and Bamenda III were also studied. Triangulation was used in data collection (questionnaires, interviews and measurements) and analysis. Findings support the hypothesis that household reproductive works that generate waste are both gendered and feminized. Gender roles and the gender division of labour make women the major agents of waste generation and management at the household level, which has far reaching implications for city sanitation. Council waste management policies and strategies, however, are not gendered even though the extent to which city managers consider gender in their policies and strategies will among other factors determine the sanitation of the city environment. The household waste generation pole and the gender dynamics within it should therefore constitute an important component that must be considered in developing municipal waste management systems that can efficiently address the daunting solid waste situation of African cities with urban environmental problems similar to Bamenda with limited resources.

List of Abbreviations

AF:	Adult Female
AM:	Adult Male
B'da:	Bamenda
BC:	Boy Child
BCC:	Bamenda City Council
BHD:	Bamenda Health District
BUC:	Bamenda Urban Council
CPDM:	Cameroon People's Democratic Movement
EWCAPAFARICA:	Environmentally Sound Design Capacity – Building for Partners and Programmes in Africa
GC:	Girl Child
GCEB:	General Certificate of Education Board
GRA:	Government Residential Area
MDGs:	Millennium Development Goals
MINATD:	Ministry of Territorial Administration and Decentralisation
MINCOMMERCE:	Ministry of Commerce
MINDU:	Ministry of Urban Development
MINEPDED:	Ministry of Environment, Forest, Nature Protection and Sustainable Development
MSWM:	Millennium Solid Waste Management
PGN:	Practical Gender Needs
SDF:	Social Democratic Front
SPSS:	Statistical Package for Social Sciences
UN – HABITAT:	United Nations Human Settlement Programme
UNDP:	United Nations Population Division
UNEP:	United Nations Environmental Programme
UNO:	United Nations Organisation
WCED:	World Commission on the Environment and Development
WHO:	World Health Organisation
WID:	Women in Development

Chapter One

General Introduction

Gender and environment concerns are key issues that have attracted global attention in recent years. The importance of these two concepts has occupied international discourses as exemplified by the World Commission on Environment and Development (WCED, 1987), Rio (1992) and the Beijing Conference (1995). Consequently, they figure prominently in the United Nations' priority concerns as advanced in the Millennium Goals (MDGs) for 2015. In the developing world, and Africa in particular, the management of urban waste constitutes one of the major environmental challenges facing their ever-growing cities. As such, concerns about environmental consciousness and behaviour of urban dwellers have taken central stage in debates on urbanisation in the continent (UNEP, 2013; UNEP, 2009a; Botkin, 2002).

The concern about waste is not only driven by the high rate at which it is generated, but also by the inability of city authorities to put in place policies, strategies, institutions and resources that can bring the situation under control. Influenced by rapid urbanisation, population and economic growth, waste generation rates outrun the management ability of urban waste managers. In the absence of much industrialization, households and markets are the main sources of municipal waste in African cities and in Cameroon. In these locations, urban waste generation is influenced by household gender politics, which is influenced by socio-cultural, economic and other factors that directly or indirectly determine gender roles. By extension, these roles influence waste generation and management, which have implications for municipal solid waste management policies and strategies that direct city hygiene and sanitation. In the paragraph that follows, we present the policy dimension of municipal solid waste management (MSWM) in the Bamenda City Council (BCC), the subject area of this study.

In Bamenda, municipal waste policies and strategies have tended to concentrate effort and resources on the collection, transportation and disposal components of waste management systems and have rarely focused on the activities and actors at sources of waste generation. Consequently, the environmental consciousness and behaviours (Dunlap & Jones, 2003) of the households and market vendors that generate much of the City's waste may not necessarily align with municipal waste strategies. Furthermore, municipal waste strategies seem not to meet waste collection or disposal needs of the urban dwellers. The result is an unclean urban environment, which has negative impacts on public health and gender roles. According to Muller & Shienberg (2003), engendering municipal waste management policies and strategies is important in making cities like Bamenda with little resources to achieve a sustainable clean urban environment. Gender mainstreamed policies and strategies require an in-depth enquiry into the gender dimension of household waste generation and management practices. In view of the centrality of the gender dimension in urban waste management, this study undertakes to make visible the principal role of women in waste generation and handling. It also seeks to identify gender gaps in the current municipal waste management system and make proposals for gender mainstreaming.

1.1 Background

Historical accounts hold that, while waste generation has always been an intrinsic part of human activities, (domestic, agricultural, industrial or commercial), waste management only became an issue of major concern with urbanization and population growth. The agricultural and industrial revolutions of the 19th century which prompted increases in production both in quantities and varieties promoted urban growth, with increased waste generation as an outcome. Limited space and lack of waste handling capability resulted in piles of waste and the associated filth and stench. Waste piles became habitats for disease transmission vectors such as rats and pests, and water supplies frequently became contaminated, perpetuating human diseases. The effects of poor waste management became evident, for example, in the Bubonic plague, which resulted from infections transmitted by flies and rodents from waste piles. These diseases ravaged the populations of Europe and Asia in the

2

mid-14th century. Whole communities were annihilated and the population of Europe was almost halved (Webster Encyclopaedia, 1977). As a result of problems like these, societies began developing waste disposal procedures, as it occurred in the ancient cities of Crete, Athens and Rome. Over the centuries, these waste disposal procedures have advanced greatly in developed countries, where despite generating a very high proportion of global waste; cities are able to combine adequate technology, human and financial resources to deal with waste disposal.

Problems of municipal waste generation amounts and management practices, in both developed and developing countries have attracted global attention. This is because of continued increase in population numbers and changes in consumption patterns (UNEP, 2013). In 2004, the United Nations Environment Programme (UNEP), in collaboration with the Global Programme of Action for the Protection of the Marine Environment from Land-based Activities, launched an awareness raising campaign focused on municipal waste management (UNEP, 2004). According to the 2007 United Nations Environmental Programme recent *Global Waste Management Market Report,* the amount of municipal solid waste (MSW) generated globally had reached 2.02 billion tonnes by 2006. UNEP estimated at that time that global municipal waste generation would rise by 37 percent by 2011, equivalent to roughly eight percent annual increase (UNEP, 2007).

Waste management is a major environmental challenge for national and local governments in developing nations (Botkin, 2002: UNEP, 2009a & b). The widely acclaimed World Commission on Environment and Development (WCED, 1987), which produced the Brandtland Report suggests that the greatest threat to water and land is haphazard waste disposal. The 1992 Rio de Janeiro Conference identifies pollution from industry and other activities, including municipal waste disposal to be a main source of environmental degradation and therefore a cause for concern. More recently, a World Health Organisation Regional Office for Africa Survey on Environmental Health covering 27 African countries prioritize MSWM second after water quality on its list of most pressing issues (Senkoro, 2003).

Economic growth, urbanization and industrialization have exacerbated the problem of solid waste disposal in developing countries by increasing the volume and changing the types of waste (Enger and Smith, 1998; Marthandan, 2007: UNEP, 2009; Buhner 2012). Relatively new waste types include hazardous waste, healthcare waste, industrial waste, chemical residues from agriculture, and, most recently, discarded electrical and electronic equipment, or E-waste. The introduction of plastic containers and bags and other components of the "throwaway society" as well as the growing volume of E-waste and waste from automobiles have compounded the problems.

The ever-increasing volume of waste requires effective waste management systems to cope with it. Waste management activities include waste storage, collection, transportation, treatment and disposal. How effectively these activities are carried out and at what costs differ among countries at various income levels, but they can be a considerable burden. The World Bank (UNEP, 2009a) estimates that municipalities in Africa and Asia spend 20 to 50 percent of available funds on solid waste management. A high proportion of council budgets in cities of developing countries are allocated for waste collection alone. Even with this high proportion, on average, 30 to 60 percent of all urban waste is left uncollected, and less than 50 percent of the population in the cities in the developing countries is served (UNEP, 2009a). These statistics represent the waste management situations of Ibadan in Nigeria, Abidjan in Cote D'Ivoire, Yaounde in Cameroon and many others (Onibokun, 1999; Afon 2007; NAPE, 2008; Sotamenou, 2010).

According to UNEP (2009b) report, waste collection alone in low-income countries drains up 80 to 90 percent of municipal waste management budgets, leaving little money for proper disposal. In middle-income countries, collection costs as much as 50 percent to 80 percent of total budget while in high income countries with much greater financial resources, it accounts for only 10 percent of the budget, allowing more funds to be allocated for waste treatment facilities. Moreover, in developed countries, like Britain and Australia, community participation, through waste separation, which is often absent in low income countries, reduces the collection costs and facilitates waste recycling and resource recovery (UNEP, 2009b).

Population growth is exacerbating problems of waste generation and management. Literature also reveals a positive correlation between waste generation, urbanization and population growth (UNEP, 2013; Tacoli, 2012; UNEP, 2009). The world's population growth rate is high. According to the United Nations, the world's population is currently above 6.5 billion and projected to gain two billion more persons by 2030. According to Smith *et al.* (2004), 95 percent of this growth occurs in developing countries, where more than 10,000 births take place every second. Tacoli (2012) and UNEP (2012) estimates that more than half the world's population now live in urban centres and that by 2050, urban dwellers probably will account for 86 percent of the population in developed countries and for 64 percent of the population in developing countries (UNPD, 2012a).

N'Dienor (2006) suggests that in 30 years, more than half of the world's population (55 percent) will be living in urban areas. According to the UN, the proportion was 30 percent in 2006 and was estimated to reach 36 percent by 2012 (UN Report, 2011). Padilla (2004) observed that in the Middle East and Asia in 1960, less than one in every three people lived in urban areas. By 2004, about 50 percent of the people were in urban centres. He projected that the proportion will rise to 70 percent by 2020. According to the UN, Africa is experiencing a very high rate of urbanization, and by the year 2020, more than half of her population will live in towns and cities. Such alarming rates of urbanization will have major repercussions for waste generation and the ability of current management systems to cope. Consequently, sanitation and waste management are likely to become even more critical (UNEP, 2013; UNEP, 2012; Conable in Dedehouanou, 1998).

As has been the situation in most countries of developing world and Africa, population growth and urbanisation have been on the increase. Urban population figures in Cameroon reveal a galloping increase from 2,202,151 in 1976, 3,994,775 in 1987 to 8,514,936 inhabitants in 2005. These figures follow a general increase in the number of towns from 195 in 1976, to 208 in 1987 and 312 towns in

2005 (BUCREP, 2010). The urban population of Cameroon in 2010 stood at 10,091,172 inhabitants as against a rural population of 9,314,928 inhabitants. The urbanisation rate increased from 28.1 percent in 1976, 37.9 percent in 1987, and 48.8 percent in 2005 to 52.0 percent in 2010 (BUCREP, 2010). This relatively high rate is associated with the creation of more administrative units. These high population growth and rapid urbanisation rates have implications for waste generation and management in cities. For this reason, regulations regarding waste generation and management have received attention.

Municipal solid waste management (MSWM) in Cameroon has been the subject of a considerable amount of legislation (Veronica, 2007). Particularly important are the laws relating to environmental management, the national environmental plan, the national water code, and the joint ministerial ban on the importation, production and use of plastics, put in place by of the Ministry of Environment, Sustainable Development and Nature Protection, and the Ministry of Commerce (Law No. 96/12, 1996; Law No. 98/15, 1998; MINEPDED/MINCOMMERCE, 2012). In addition, several ministerial departments in Cameroon have the mandate to implement solid waste management regulations. These include departments in the Ministries of Territorial Administration and Decentralisation; Environment, Sustainable Development and Nature Protection; Urban Affairs; Town Planning and Housing, Public Health; Industrial and Commercial Development; and Finance and the Budget.

The national average solid waste generation amount per capita per day in Cameroon is about 0.60 kg in the major cities of Douala, Yaounde, Bafoussam, Bamenda and Garoua. The national average daily total is estimated at 12,000 tonnes per city. The organic waste component though declining is far greater than the inorganic waste material fraction which is promoted by changing lifestyles and consumption patterns (Sotamenou, 2010; Ngnikam & Tanawa; 2006; Achankeng 2005). As in most cities of the developing world, waste management in Cameroon lags behind waste generation. Consequently, waste is often found littering roadsides, markets, and street corners, clogging drainage channels, and posing health and environmental threats. Regional capital cities, and particularly the capital cities of Yaounde and Douala, vividly exemplify most of these problems (Sotamenou, 2010; Ngnikam & Tanawa; 2006; Achankeng 2005; Fombe, 2005).

The city of Bamenda, the subject of this study, shares the waste generation and management experience of many towns in Cameroon. Solid waste management in the City Council of Bamenda is a public sector service. The responsibility for conceiving and implementing waste management policies and strategies and activities rests with the council authorities of the three communes within Bamenda; BAMENDA I, II, and III, under the supervision of the city council. Collaboration from related government departments such as Urban Development and Housing and the inhabitants is expected.

Municipal solid waste management (MSWM) in Bamenda has evolved alongside the evolution of the city in its history from colonial days to the present. During the colonial period, waste management in Bamenda, with a focus on urban hygiene and sanitation was the responsibility of the Mankon Area Council under the supervision of the medical doctor in charge of preventive medicine (Chuba, 1978). Waste bins were provided by the council to collect waste for evacuation to incinerators. Incineration was the main waste disposal technique used by the council. With population increase, the incineration sites became too close for comfort, as inhabitants complained that the fire and smoke from incineration sites were dangerous (ibid). The waste bins were also sources of bad smells, flies and rodents. There was clearly a need for a new technique. The council then opted for collection and evacuation of waste away from home environments. On April 9th 1970, the Mankon Area Council purchased its first tip-up truck to remove waste from its own part of the town (BUC File, No. 319 Vol.1/68). Households were called upon to transfer their waste to specified collection sites for onward evacuation and disposal. Farmers who were interested were asked to apply for council to dispose of waste on their farms. Initially, this service was free but later delivered at a hired charge of 100FCFA per kilometre paid by the desired farmers. The medical doctor in charge of preventive medicine drew up and made public the waste collection itinerary for the town for the week. The areas concerned were:

Community Layout and Small Mankon Areas; Nta Mbag and environs; and Hospital and Atuakom area. In spite of these efforts, large quantities of solid waste remained uncollected. However, waste management at this time could be judged to be relatively more effective than now with much concern for public health. This could be expected as the population numbers were relatively less than today and consumption and lifestyles presented few variations.

With the reunification of West and East Cameroon in 1972, Bamenda gained the status of being first, the divisional and later the provincial capital of the Mezam Division and the North West Province. This new status led to an influx of population to serve in the new administrative departments and auxiliary services, which increased the volume of solid waste. Unfortunately, the ability of the council to handle the waste increase lagged behind the rate of waste generation. The consequences are highlighted in the governor's letter to the council in 1973 (BUC File, No. 319 Vol.1/68). There he expressed disappointment with the state of waste management for the town, which had not improved alongside the new status of the town. The issues he raised included the following: Domestic animals killed by passing vehicles are left to decompose in the streets for days; stagnant water and refuse heaps are allowed to breed mosquitoes, rodents, and dangerous reptiles; market places are always dirty looking and lawns and hedges are hardly or never cared for (BUC File, No. 319 Vol.1/68).

In 1980, there were attempts to improve on waste equipment and services (Nso, 1998). A tractor, chassis and container were added to the lone and aging tractor, which was becoming increasingly ineffective due to constant breakdowns and repairs. In addition, metal waste collection bins were placed at strategic locations in the town. Two official dumpsites were created, one at the Old Fish Pond (which now serves as the Bamenda City Council Food Market) and the other near Government Bilingual High School (GBHS) at Ntamulung. The monthly clean-up campaign became another initiative. Until 2013, the council, in collaboration with the Senior Divisional Officer, the gendarmerie and health services, selected one day a month for general clean up, referred to as the "clean-up campaign" day. On that day, inhabitants of the town did a thorough cleaning of their immediate environments (home and work places). The waste generated was assembled in piles at roadsides pending collection by the council waste van on that day or the day after. This practice was suspended in 2013 when the City Council contracted private waste operators to sweep and grub gutters in the city.

Currently, the situation of municipal solid waste management for the city of Bamenda remains wanting. Unprecedented urbanisation and economic growth accompanied by heterogeneous populations and dynamic lifestyles have caused increases in waste quantities and material composition of particularly non-biodegradable waste materials. Unfortunately, the current municipal waste management system is unable to cope with these changes, and the rate of waste collection and disposal continues to lag behind waste generation. The consequences are heaps of waste at roadsides and in gutters, valleys, streams and rivers that spoil urban aesthetics, harm the environment, and endanger public health (Anschutz *et al.* (1995), Nso (1994), Balgah, 2002 and Achankeng (2005).

Scholars like Nsoh (1994), Balgah (2002) and Achankeng (2005) have advanced reasons for the inadequacy of municipal solid waste management in Bamenda. Inappropriate construction of houses and poor urban planning render accessibility into the interior parts of neighbourhoods difficult for the council's waste collection vans. Irresponsible waste disposal practices of urban dwellers do not comply with council waste regulations. Waste management technology and the equipment available are insufficient, and there are constant breakdowns of existing equipment that need repairs and make waste collection burdensome. There is also a shortage of qualified staff for repairs of waste equipment and insufficient funds to purchase more waste equipment and ensure the running cost of waste service delivery. In addition, the level of implementation of existing environmental and waste policies by the council is weak (Anschutz *et al.*, 1995; Nso, 1994; Balgah, 2002; Achankeng, 2005).

Actors involved in waste generation and management activities especially at the primary level (as households and market places) are considered to play a very important role in the success of any waste management system instituted by city authorities (Enger & Smith, 1998). Household gender division of labour attributes different roles to women and men. In over 80 percent of households in the developing world, women perform household reproductive works, i.e., unpaid work necessary to maintain the household such as domestic chores and care provision. According to Elson (2000), this unpaid domestic work promotes gender inequality as these tasks are traditionally performed by women and taken for granted by most governments. Men, for their part, rarely share in unpaid domestic chores, as is the case in western countries (Baxter, 2002). Moreover, even women who carry out paid work in the public sphere still come home to do reproductive work, and even in the case where domestic servants are hired, most often they are female. In short, women are highly involved in domestic chores that have the potential to generate waste. These activities include food preparation, sweeping and clearing of the house and immediate home surroundings. Furthermore, household waste handling activities, including storage, sorting, composting, reuse, transportation and disposal, are also gendered. The gender of the actors involved and their practices have implications for the implementation of council waste management strategies and for the hygiene and sanitation status of the city.

Owing to traditional roles especially in countries of Africa, women in households perform reproductive roles of child bearing/upbringing and managers of the domestic environment (Moser, 1993) that bear on waste generation and management. The United Nations Organization (UNO) in its Millennium Development Goal (MDG) number three advocates gender equality and the empowerment of women, and the Commonwealth Secretariat argues for a participatory approach in development projects and programmes, with women as important partners. In addition, and especially relevant to this research, the United Nations Human Settlement Programme (UN-HABITAT, 2008) argues the case for gender considerations in municipal council service provisions. The important role of women in solving development problems has also been acknowledged in Cameroon. By the creation and organisation of the Ministry of Women's Empowerment and the Family with the goal of empowering women, the Cameroon government has recognised the role of women in the development agenda of the state. Councils, too, need to recognise women as they seek to achieve clean urban environments. Given that women make up some 52 percent of the population in most countries, local democracy, inclusiveness and sustainability can only be achieved when this segment of the population has an equal say in the way that cities and municipalities are organized and managed (UN-HABITAT, 2008). More specifically, this means that women must be involved in decision-making at all levels so that their needs and priorities are also reflected in urban planning and municipal service provision, including waste management. Improvements in waste technology and resources are not enough to achieve efficiency in waste management that can guarantee a sustainable clean environment. There is the urgent need to mainstream gender in the municipal waste management system in Bamenda and other cities with similar characteristics.

1.2 Statement of the Problem

The management of urban waste constitutes one of the major environmental challenges facing African cities. As such, concerns about environmental consciousness and behaviour have taken central stage in debates on urbanisation in the continent. Accompanying rapid urbanisation and unprecedented population growth, is the solid waste generation rate, which supersedes the capacity of city authorities to handle efficiently. Consequently, huge mounds of refuse are very visible along street sides, streams, culverts, and even very close to dwellings. The devastating effects of this poor waste disposal include degraded aesthetics, inconvenience, environmental pollution, and major risks to public health for cities and inhabitants (UNEP, 2013; Tacoli, 2012; UNEP, 2009; Marthandan, 2007; Botkin, 2000).

Undesirable urban behaviours accompanied by councils' inabilities to sufficiently provide social amenities have negatively compounded the plight of the urban poor (especially women) as portrayed by the effects of the consequences waste, air and water pollution in the overcrowded urban cities. Thus the management of urban waste by the actors involved must begin with a mastery of waste generation and handling practices; particularly domestic waste and market waste which accounts for more than 70 percent of all waste in most African cities (Afon, 2007; Achankeng, 2005).

Failure of MSWM systems in Bamenda, like in most other cities in Africa, has been attributed to inadequate finance, inadequate technology, lack of expertise and other resources, and poor planning (Achankeng, 2005; Balgah, 2002). This thesis argues that the omission of a gender perspective in MSWM policies and strategies that takes into consideration the role of men, women and children is another major barrier to efficient waste management. At present, household waste generation and management actors and activities are gendered, but council waste management policies and practices are not. Therefore, the Bamenda City Council's approach to waste collection and management cannot be effective if it neglects the gender perspective, and the provision of adequate technology and resources alone will not guarantee efficient waste management. Rather, waste governance policies and strategies that take into consideration the different sources of urban waste generation, different actors involved in waste generation and management, and the interest of those implicated are needed. In this light, this study investigates the actors and resources involved in waste generation and management within the Bamenda City Council bringing out the central role of women. The study further discusses how mainstreaming gender in council waste policy and strategies can bring effective and sustainable management. That is, making men, women and children's concerns and experiences an integral part of MSWM plan. Against this backdrop, the following research questions are posed.

1. To what extent does the Bamenda City Council consider the central role of women in household waste generation and handling in its policy and strategy of urban waste collection and disposal;

2. Are there gender differentials in the level of environmental consciousness and behaviour in men and women's perceptions and management of household waste?

3. Can waste generation in markets in Bamenda be characterised along gender lines?

4. To what extent are Council resources and waste delivery strategies responsive to inhabitants' waste management needs in general and to women's gender needs in particular?

5. What gender-sensitive sustainable waste management policy and strategies can be employed by the Bamenda city council that will address current deficiencies and improve urban livelihood and sanitation?

1.3 Objectives

1.3.1 General Objective

The main objective of this study is to analyse gender issues and gender gaps in municipal solid waste generation and management within the Bamenda City Council and recommend options for mainstreaming gender in municipal waste management policies and strategies.

1.3.2 Specific Objectives

In order to attain the general objective, the following specific objectives have been set:

1. Describe the demographic profile of households and market waste actors within the Bamenda City Council

2. Examine household gender reproductive roles and their potential to feminise household waste generation

3. Examine household gender reproductive roles and their potential to feminise household solid waste management practices

4. Examine market waste generation potentials, the characteristics of the waste and their implications on municipal solid waste management.

5. Determine the extent to which municipal solid waste collection strategies influence households' and market vendors' waste generation and handling practices and their implications for effective management;

6. Examine the effects of current municipal solid waste disposal policies and strategies on the environment and implications for women's gender roles;

7.   Make recommendations for mainstreaming gender in MSWM to enhance efficient management of urban waste in Cameroon cities

1.4 Justification and Significance of Study

A survey carried out by Hart *et al.* (1995) on environmental problems in the city of Bamenda ranked municipal waste management and drainage as Bamenda's most pressing problems. Similarly, the Ministry of Urban Development's (2011) in its diagnostic report on environmental problems pointed at waste management as a core cause of air, water and land pollution and poor drainage in Bamenda (MINDU, 2011). Markham & Fonjong (2014) researching on environmental problems, identify waste disposal as one of the environmental problems facing Cameroon and definitely Bamenda. These studies suggest that the problem of MSWM is endemic, real and requires urgent attention. Recent cholera epidemics in several cities in Cameroon, notably Yaoundé, Maroua, Limbe, Buea and the administrative unit of Bafut subdivision in Mezam Division, also suggest that research on waste management is of high priority, as poor disposal and poor sanitation serve as vectors for cholera transmission.

Waste is an inconvenience, but it is equally a resource, which if properly harnessed, can reduce council expenditures and raise revenue for the council and its citizens. Waste collection and recycling can also serve as a source of livelihood for a population with a high unemployment rate. Moreover, in a municipality where household and market waste constitute more than 90 percent of total municipal solid waste (Achankeng, 2005), and where waste generation and management is steered by gender division of labour and gender power relations, mainstreaming gender in waste management strategies and techniques increases the chances for effective sustainable management.

This research is therefore timely because it is in line with the now widely accepted view that incorporating a gender perspective in development efforts is necessary for their successful implementation. The study emphasizes the role of gender in municipal waste

14

generation and management, and informs councils about how to develop waste management strategies that are sustainable and gender inclusive. The study is also relevant to current global concerns as it addresses two Millennium Development Goals relating to the degradation of the environment, gender equality and women's empowerment.

The literature review, summarized in the next chapter, reveals that waste management in Cameroon and Bamenda in particular, has been the subject of little academic research. Indeed, very little comprehensive data on waste generation quantities and characteristics and their management, not to mention gender and waste, exists. These inadequacies provide a raison d'être for the present study, which it is hoped, will contribute to the database. More so, a study on waste management is important at this moment when the government of Cameroon expresses great concern about environmental deterioration. This concern has been expressed by the recent ban on the production, importation and use of plastics.

A study of this nature that highlights the different sources of waste, types and quantities of waste and the actors involved, will provide useful data to waste management agents, both public and private, and particularly the BCC and the Regional Delegation, Nature Protection and Sustainable Development of the Environment that have the statutory responsibility to manage waste. The findings will identify weaknesses of the current waste delivery service and make recommendations for policies and strategies that can enhance the system. Specifically, they will guide these officials in selecting improved waste collection strategies, an appropriate labour force, and an optimal schedule for collecting municipal waste and understanding the waste resource potentials of the municipality. For the community, the study results will help to raise environmental awareness about the effects of poor waste disposal practices.

1.5 Delimitation of the Study

This study is limited to the urban zones of the sub-divisional councils of Bamenda I, Bamenda II and Bamenda III for which waste collection and transportation is controlled by the BCC management.

1.6 Limitations of the Study

Most of the data for the study come from a household survey. Data collection through the administration of questionnaires was flawed by the high non-return rate in the health district area of Atuakom. In this district, self-administration was anticipated by the research assistant who was then unable to retrieve completed questionnaires. The researcher is of the opinion that the low return rate did not distort the findings or reduce the authenticity of information from households of the area. This is because the situation was resolved by purposefully identifying and selecting a quarter (Abangohh), within the Atuakom health district neighbourhood for on-site waste analysis.

1.7 Operational Definition of Terms

Gender. This is the socially constructed roles and relationships, personality traits, attitudes, behaviours, values, relative power and influence that society ascribes to male and female on a differential basis. It is an acquired identity that is learned, changes over time, varies widely across and within cultures (Pearson, 2006; Woroniuk *et al.,* 1997). In operational terms, gender is used to refer to the roles of adult women, girl children, adult men and boy children in the household.

A *gender analysis* is a systematic examination of the differential impacts of economic, political and social policies, programmes and legislation on women and men. A gender analysis is used to identify, understand and describe gender differences, interests, contributions and the impact of gender inequalities on a sector or programme (USAID, 2010; Pearson, 2006). In operational terms, gender analysis in this study focuses on the level of participation of the adult male, adult female, girl child and boy child in household and market activities with potential to generate and manage waste.

Gender mainstreaming: It is the process of making women's as well as men's concerns and experiences an integral dimension in the design, implementation, monitoring and evaluation of policies and programs in all political, economic and social spheres, so that inequality between men and women is not perpetuated (Pearson, 2006).

Gender needs/interests are necessities, concerns and values of persons that result from their sex, age, class, etc. Gender needs are divided into two categories: Practical Gender Needs and Strategic Gender Needs (Moser, 1993; Visvanathan *et al.*, 1997; March *et al.*, 1999; Taylor, 2001).

Practical Gender Needs (PGNs) are necessities of life, which improve the condition of women and men in the short term. Actions that address women's practical needs are oriented towards welfare, health care, provision of food, etc. These are needs, which are identified with women as wives, mothers and caregivers. In this study, practical gender needs include the waste collection and disposal services placed at the disposal of households by the BCC and other waste management agents (Moser cited 1993: 57).

Strategic Gender Needs (SGNs) are long term changes needed to improve women's subordinate position to men. They are intended to challenge unequal relations between men and women. For instance, making waste recovery at the level of the household an income generating activity would financially empower women and change household gender power relations to the benefit of women (Moser cited in March, 1999:58).

Productive Work: According to the European Commission cited in Schneider (2006), productive work is making goods and services for which cash value is attached. Men and women are involved in productive activities, but their functions and responsibilities often differ. Women's productive work is often less visible and less valued than men's work.

Reproductive work is care and maintenance of household and household members through such activities as child bearing and caring for children, preparing food, collecting water and fuel, shopping, housekeeping and family health care. According to Caroline Moser (1993), reproductive work in poor communities is generally labour intensive and time consuming. Reproductive work is often carried out by women and girl children. Waste handling activities at the level of the household for which remuneration in monetary value is not attached is reproductive work.

17

Ecofeminism (ecological feminism) is the position that there are important connections in the way one treats women, people of colour, and the underclass on one hand and how one treats the nonhuman natural environment on the other (Warren, 1997).

1.8 Chapter Presentation of Thesis

This thesis is presented in eight chapters. Chapter one is a general introduction comprising the background of the research, statement of the problem, research questions and objectives. It also presents the significance, limitations and delimitations of the study, and the definitions of some important terms and concepts used. Chapter two reviews literature and discourses on waste generation and management in the developed and developing world, Cameroon and Bamenda. Chapter three dwells on the area of study and the research methodology employed. In chapters four, five, six and seven are presented the research findings addressing the different objectives of the study. In chapter eight, discussions, conclusions from findings and recommendations for policy and further research are made.

Chapter Two

Literature Review

For waste management plans and strategies to be efficient and sustainable, a solid information base about waste generation and waste management activities and the stakeholders involved is vital. This chapter focuses on literature related to waste generation activities, quantities, types and sources. It further looks at waste management activities and techniques in developed and developing countries, in Cameroon, and in the city of Bamenda. In addition, this chapter reviews information about the role of gender, and particularly that of women, in municipal waste generation and management. Before engaging in literature on waste generation and management, an understanding of what waste is perceived to be is of relevance.

2.1 The Concept of Waste

'Waste' has been variously defined by governmental bodies depending on the context. According to the United Nations (UNEP, 2009), waste refers to materials that are not prime products meant for consumption or further use. This implies that wastes are not products directly produced for the market; they are products for which the initial user has no further use in terms of his/her own purposes of production, transformation or consumption and which he/she wants to dispose of. The Basel Convention defines waste as substances or objects which are disposed of or intended to be disposed of or are required to be disposed of by the provisions of the national law. Under its Waste Framework Directive, the European Union defines waste as an object that the holder discards, intends to discard or is required to discard. The Cameroon Environmental Law No. 96/12 of 5th August 1996 on its part defines waste as any residue from a production process or, any movable and immovable goods abandoned or to be abandoned.

Scholars have also attempted to define waste. Brown (1991) defines it as superfluous refuse that no longer serves a purpose and is left over after use. He also refers to it as a useless by-product of manufacturing or physiological processes. Eleanor *et al.* (1998) describe refuse as including any movable material that is perceived to be of no further use and is permanently discarded. Micheals (1966 cited in Eleanor 1998) suggests that waste could be classified as solid, liquid and gaseous and notes that the physical nature can change with conveyance or treatment. In light of all these varying definitions, it is difficult to consider anything as waste in some absolute sense; however, in the context of this study, the term 'waste' is used to refer to any discarded material not deemed any longer useful to the owner for the primary reason for which it was initially acquired. Some waste, but not all, could continue to be used, either by the owner or another person.

2.2 Municipal Solid Waste Generation and management in the Developed World

2.2.1 Municipal Solid Waste Generation in Developed Countries

Worldwide, waste generation is on the rise. The Developed Countries of the North, including countries of the Organization of Economic Co-operation and Development) (OECD), the European Union, Canada, United States of America, Russia and Japan, are the most technologically advanced areas of the world. These regions and countries display high levels of income, affluent lifestyles and high levels of consumption. These are characteristics that tend to encourage waste generation. For example, according to statistics from the International Bank for Reconstruction and Development and the World Bank, these countries in 1999, represented only 16 percent of the world's population but consumed 75 percent of global paper production (IBRD and World Bank 1999). Municipal solid waste generation in these societies is prolific, as their populations have the tendency to 'just throw away' things after using them (US EPA, 2003; Botkin, 2001; Newton &CSIRO, 2001; Wright, 2000). In general, municipal solid waste generation continues to rise in absolute and per capita terms in the industrialized countries (Table 2.1).

Table 2.1: Waste production in some countries of the Developed World (2005)

Country	Production (per capital/Kg/day)	Country	Production (per capital/Kg/day)
USA	2.07	Sweden	1.02
Canada	2.00	France	1.00
Finland	1.71	United Kingdom	0.95
Australia	1.70	Belgium	0.94
Netherlands	1.36	Germany	0.91
Norway	1.29	Turkey	0.90
Switzerland	1.20	Spain	0.88
Japan	1.11	Greece	0.81
Australia	1.04	Portugal	0.70
Russia	0.43		

Source: Adapted from Sotamenou (2005, 2010))

Table 2.1 above presents estimates of daily waste production in some countries of the developed world. The figures indicate that while waste production is generally high, countries at about the same level of development differ greatly in their waste generation rates. For instance, USA and Canada are the world's leading countries of waste generation, with 2.07 and 2.00 per kilogram per capita per day. This is closely followed by European countries of Finland, Netherlands, Norway and Australia. Countries with relatively lower amounts include Russia (0.43), Portugal (0.70) and Greece (0.81).

2.2.2 Municipal Solid Waste Management in Developed Countries

In this section, literature is reviewed on waste composition, actors and stakeholders in urban waste management strategies and disposal techniques.

2.2.2.1 Waste Composition

Michaels (1996) produced a useful classification of municipal solid waste. Municipal solid waste is comprised of rubbish, ash, bulky waste, street refuse, dead animals, abandoned vehicles, construction and demolition wastes, industrial refuse, special wastes, (that, not classified), animal and agricultural wastes, and sewage treatment residue. Waste composition in industrialized nations tends to contain more non-biodegradable material than biodegradable as opposed to the developing world (OECD, 2008; Kapepula, 2006; Damodaran *et al.*, 2003; Ngnikam, 2000). For example, according to USEPA (2003), household solid waste in the USA contains the following components: paper (38 percent), yard trimmings (12 percent), food waste (11 percent), plastic (10 percent), metal (08 percent), glass (05.5 percent), wood (05.3 percent) and others (05.3 percent). According to OECD (2008), Canada was ranked first in the production of paper waste (47 percent).

2.2.2.2 Actors and Stakeholders in Urban MSWM

Actors and stakeholders (individuals, agencies, non-governmental organizations, councils, etc.) in urban municipal waste management are generally involved in waste generation and management activities of sorting and separation, waste collection and removal, waste reduction, waste recovery and waste disposal. Related tasks include street sweeping, clearing and repairing drains, transforming and reusing discarded articles, and supplying waste collection equipment at waste generation centres (USEPA, 2003; Onibokun and Kumuyi, 1999). Literature identifies two main actors active in formal waste management in urban municipalities, the public sector and the private sector.

Public participation in solid waste management by national and local government occurs in two ways. First, the state plays a leading role in setting forth laws governing waste management and executing them. Second, government may actually manage waste collection and disposal. In the US, for example, the state governments develop the legislation and policy guidelines regarding waste management, and local government does most of the waste collection and disposal. In other countries, the government develops the legislation and policy guidelines, but allows the private sector to execute them. For example, in Canada, the private sector takes the leading role in collection and transportation of municipal waste (US EPA, 2003).

2.2.2.3 Municipal Solid Waste Management Strategies and Techniques

Increasingly MSW generation necessitates management strategies and techniques that are sustainable, cost effective and adapted to the local environment. Waste management strategies and techniques have been developed and used in relation with the type of waste activity to be handled. These activities are waste collection and storage, transportation, treatment and disposal (Onibokun, 1999; Thu, 2005; UNEP, 2009a). In most countries of the developed world, on-site waste streaming (separation), collection and storage by households, establishments and industries follow legislated rules. Standardized containers designed with environmental safety, ease of recycling, and collection technology in mind are used (Newton *et al.,* 2001; Hansen *et al,* 2002; Porter, 2004; Achankeng, 2005). Waste treatment is done after waste has been collected. Waste treatment refers to activities intended to facilitate recycling, resource recovery, composting, transportation and disposal of solid waste. Some of these activities include bailing, shredding and grinding to reduce volume.

2.2.2.4 Waste Disposal Techniques

Waste disposal, along with recycling and resource recovery, is one of the outcomes of waste collection and waste treatment. International standards demand that the residues of waste treatment be disposed of in sanitary landfills or incinerators (IBRD and the World Bank, 1999); however, the importance of landfills varies among countries. According to OECD, the tendency is for countries with high population densities, such as Japan, the Netherlands, the UK and Switzerland, to adopt incineration as the major solid waste management method. UNEP (2000b) notes that incineration of municipal solid waste is operating successfully in the emerging industrialised countries of South Korea, Taiwan, Hong Kong, Indonesia and Singapore. The countries of Hong Kong, Indonesia and Singapore incinerate about 90 percent of their waste (UNEP, 2000b). Low population density developed countries, such as Australia, Canada and the US, exhibit high rates of waste landfilling (OECD, 2004). According to US EPA (2003), the US generated 230 million tonnes of MSW in 1999. Of this amount 57 percent was landfilled, 15 percent burnt and 28 percent recycled. UNEP (2000) points out that fewer but larger and more environmentally friendly regional landfills, called 'mega fills', are now preferred in the US. In Canada, of the 29 million tons of MSW generated in 1998, 90 to 95 percent was either land filled or incinerated. The rest was diverted through recycling, composting or reuse.

Landfills, however, are far from a perfect technology. Poorly constructed or poorly managed landfills are a threat to the environment and public health due to the gases and leachate that emanate from them. In developed countries, stringent regulations on landfill designs, construction and execution are in use, with the aim of making landfills more environmentally friendly. These include leachate management, which entails putting adequate liners to contain and direct leachate to treatment plants before discharging it to the environment and the venting or collection of landfill gases for the generation of electricity. Furthermore, provisions must be made for the closure and post closure monitoring and restoration of landfill areas (OECD, 2004; OECD, 2008).

Fortunately, the environmental impacts of waste disposal have exhibited a downward trend in OECD countries in recent years. UNEP (2000c) attributes this decline to the institution and enforcement of extensive regulations, especially landfill and incineration standards, which have been accompanied by the development of highly efficient technologies. In addition, advanced incineration technologies significantly reduce the emissions of particulates, nitrogen oxide, sulphur dioxide, carbon monoxide, heavy metals, and dioxins. It is however important to note that, the long term effects of contamination of surface and ground water and the soil due to earlier poorly managed landfills remains problematic.

2.3 Municipal Solid Waste Generation and Management in the Developing World

2.3.1 Municipal Waste Generation in Developing Countries
In the developing world, which comprises most nations in Africa, Asia, Latin America and the Caribbean, and Oceania excluding

Australia and New Zealand, the major driving force behind waste increases has been the high rate of population growth and urbanisation (United Nations Population Division, 2002). Fast growing populations lead to higher rates of consumption and invariably more waste generation. Table 2.2 present daily waste generation amounts per capita for some selected countries in the developing world.

Table 2.2: Waste Production in selected Low Income Countries (2005)

Country	Production (per capital/Kg/day)	Country	Production (per capital/Kg/day)
Mexico	0.98	Brazil	0.51
Malaysia	0.75	India	0.50
Morocco	.075	Niger	0.50
Burkina Faso	0.61	Pakistan	0.50
Cameroon	0.60	Philippines	0.50
Nigeria	0.60	Ghana	0.40
Senegal	0.60	Vietnam	0.40
Tchad	0.60	Mauritania	0.35
Egypt	0.58	Tanzania	0.36
Bolivia	0.56	Indonesia	0.30

Source: Adapted from Sotamenou (2005, 2010) and Achankeng (2003, 2005)

Table 2.2 reveals that with only a few exceptions, waste generation per capita per day is less than a kilogram a day for most developing countries, less than developed countries (Table 2.1), where waste production rates are usually more than one kilogram per capita per day. This relatively lower rate can be attributed to low standards of living and poverty. Nevertheless, there is an increasing variety in the waste composition materials in developing countries with increasing globalization (Achankeng, 2003). The importation of relatively cheap manufactured goods from the new industrialising nations of South East Asia and Latin America has encouraged the generation of new waste types, such as E-Waste materials.

According to the United Nations Environment Programme and the Environmentally Sound Design and Management Capacity-building for Partners and Programmes in Africa, average waste generation rates in African cities range between 0.50 and 0.87 kg per capita per day (UNEP, 2000a: ENCAPAFRICA, 2004), while Asian cities experience a very wide range of MSW generation depending on the level of industrialization and consumption patterns. The IBRD and World Bank Report (1999) reported that per capita per day waste generation rates in Asian cities vary from an average of .64kg among low-income countries, through 0.7kg in middle-income, to 1.1kg in high-income countries. India and Hong Kong present on average extreme rates of 0.35kg and 5.0kg respectively (Buekering *et al.* 1999). MSW generation rates in Latin America and the Caribbean range between 0.3 to 1kg per capita per day (UNEP, 2000d).

2.3.2 MSW Management in Developing Countries

Municipal solid waste composition in the developing world is dominated by biodegradable material, which comprises 70 to 90 percent of MSW in Africa (Furedy, 2002: ENCAPAFRICA, 2004; Cofie *et al.*, 2010) and 30 to 75 percent in Asia and South America (Lardinois and Marchand, 2000; Behmanesh, 2010). This higher proportion of organic material in the waste can be attributed to the nutritional habits of countries of the south, which includes much reliance on vegetables and fruits and less on pre-packaged products, some of which come from home gardening and urban farming (Kapepula, 2006; Tech-Dev, 2003; www.environcorp.com).

2.3.2.1 Actors and Stakeholders in Urban Municipal Waste Management

As is the case with developed countries, public sector involvement in waste management is concerned with policy and in some cases actual involvement in waste collection and disposal. The public actors who include government departments such as the departments of health, city planning, and local councils provide waste management regulations (Kironde, 1999; Onibukun and Kumuyi, 1999). The participation of the private sector in municipal solid waste management in Africa in recent time has become very active. Waste management agents in the private sector include companies and individuals who handle activities of waste management, with waste collection and transportation being the most common. Municipal waste collection by contract is becoming increasingly common in many developing countries of Asia and Africa (Dedehouanou, 1998; Thu, 2005). This is the case with Mulinet in Dar es Salaam, Tanzania, Ibadan Urban Sanitation Committee (IUSC) in Nigeria and Cameroon (Kironde, 1999; Onibukun and Kumuyi, 1999; Sotamenou, 2010). In Cameroon, some city councils take the full responsibility to handle waste from collection to disposal while others have turned to the private sector.

In much of the developing world, "official" public and private waste management in the formal sector operates alongside an informal sector. Waste management activities within these sectors are handled by different actors working either in parallel or in collaboration. In the formal sector, municipal councils usually have the statutory responsibility to manage municipal solid waste. Other actors include the national government authorities, external support agencies (ESAs), Community Based Organisations (CBOs), contractors and Non-Governmental Organisations (NGOs). The role of NGOs in addressing waste issues and environmental problems is well documented. In this regard, Markham and Fonjong (2008, 2012 & 2014) have studied NGOs and the role of NGOs as environmental movements in addressing environmental problems in the developing world in general and Cameroon in particular. Findings suggest that NGOs and their activities as environmental movements are limited and infrequent unlike in developed countries like Germany (Markham, 2008).

These studies argue that the mobilization of citizens for confrontation with the government or business interest is rare. Reasons advanced for this slow or lack of activism are linked to the fact that, the people's livelihoods and health have not been sufficiently threatened. Other factors include a resilient, quasi-authoritarian government characterized by patronage, divide and conquer strategies, and mild repression. The studies however conclude that there exist some NGOs and citizens' groups in Cameroon that address environmental problems through lobbying, public education, and concrete projects to protect the environment. Some of such NGOs in Bamenda include COMINSUD and Paradise on Earth. Studies recommend that deeper thoughts be given to environmental issues and environmental movements (NGOs and CSOs) in order to enhance environmental protection.

Other studies view scavengers and Scavengers' Cooperatives, households and individuals as the main actors in the informal sector (KIRONDE, 1999; Ali *et al*, as cited in Achankeng, 2005: UNEP, 2009; Markham & Fonjong 2008). For example, according to Onibokun (1999), Attahi (1999), and Afon (2007), who studied waste management in the cities of Ibadan and Ogbomoso in Nigeria, and Abidjan in Cote D'Ivoire, small waste operators, use wheelbarrows to collect waste over short distances. They function in markets, motor parks and in residential areas and charge their clients agreed sums of money. They dispose their collected refuse in depots within the neighbourhoods pending onward collection and transportation by the formal waste agents.

In spite of the efforts of both the private and public sectors, effective waste management and its cost remain daunting problems for most cities in developing countries. The failures of waste management efforts are attributed to several factors identified by Dedehouanou (1998), Achankeng (2005), Sotamenou (2010), Ngnikam (2006) and Balgah (2002), which carried out waste studies in Cotonou, Yaounde, Bamenda, and Douala respectively. The drawbacks include persistent government failure in coordinating different actors and activities, failure to supply supportive infrastructure, failure to institute a real dialogue between all stakeholders, insufficient funding and inadequate technology and equipment. Dedeouanou (1998) suggests that more government intervention is needed in cities like Cotonou to attain waste delivery service provision that is effective and efficient. Like the other researchers, he adds that government and municipal waste management can be improved if the following conditions can be attained: reducing population pressure, improving the performance of organizations in charge of home waste management, raising people's awareness about the environment and alleviating poverty through employment (Dedehouanou, 1998; Onibokun and Kumuyi, 1999). Thu (2005) argues that such conditions can only be very successful if gender is mainstreamed in each of these conditions.

Limited knowledge of local realities of developing countries tends to dispose donor agencies to transfer technologies used in home countries to recipient communities. In some instances, obsolete and outdated equipment are offered as foreign aid, which in the long term turns into liabilities rather than assets. For example, the automated equipment offered to the Bamenda council with all good intent and purpose, now poses a major financial burden as repairs and maintenance are required constantly, and the expertise and spare parts are absent. Mensah and Larbi identify ill-adapted technology as one of the waste management problems in Ghana. These researchers (Mensah and Larbi, 2005 cited in Buhner, 2012) propose that the Ghanaian experience within a weak socioeconomic context, will be better off with a manual waste management system.

In an attempt to propose solutions to the waste management problems in Benin City and other urban areas in Africa, Dedehouanou (1998) suggests that the relationships among government (municipality), urban dwellers, private firms, NGOs, self-help organizations, and individuals, influence the effectiveness of waste management systems (policies, strategies and techniques). A participatory approach to waste management, which requires the collaboration of all stakeholders in the process of waste management in an effort to reach a comprehensive and desirable outcome, is frequently recommended (Onibokun, 1999; Attahi, 1999; Thu, 2005, Cofie *et al*. 2010).

Cofie *et al*. (2010), for example, recommend multi-stakeholder consultation, in which joint planning and joint decision-making precede adoption or modification of existing waste management policies, strategies or techniques. In developing countries like Cameroon, where industrialization is minimal and large-scale waste management for resource recovery through recycling is difficult to implement, policies should be oriented towards reuse. In such a situation, the key stakeholders whose opinions are imperative will be the small and micro enterprises and households where most waste is generated (Olufunke Cofie, Rene Van Veenhuizen, Verele de Vreede, Stan Maessen, 2010). Along the same lines, a non-governmental organisation, Integrated Support for Sustainable Urban Environment (ISSUE, 2007-2010), suggests that genuine decentralized waste management is best done by a programme board accompanied by district level field management.

2.3.2.2 Waste management strategies and techniques

In most low-income countries, waste storage methods at the level of the household are not standardized. In a majority of African cities for example, storage methods and equipment include drums, plastics bags, buckets, plastic pails, paper, and bamboo baskets (Onibokun, 1999). In most cities in Africa, waste generation far exceeds waste collection. According to UNEP – IETC (1996), only about 40 to 50 percent of MSW is collected.

A variety of transportation equipment is used to move waste from generation sources to collection and disposal sites. The equipment ranges from locally adapted equipment, such as handcarts, wheelbarrows, tricycles and pushcarts, to conventional open trucks, side and rear compactors, and trailers. UNEP-IETC (1996), states that 70 percent of waste in West Africa is transported with the use of motorized equipment. The equipment used is often poorly adapted to the high density of bio-waste, the terrain, the moist climate, and narrow and unpaved roads, making mechanical breakdowns common and curb side collection impractical (Medina, 2000). The transfer system for moving refuse to dumps and landfill sites varies across countries. However the most common system as is practiced in the cities of Abidjan in Cote d'Ivoire, Lagos in Nigeria, Kampala in Uganda, Yaounde and Douala in Cameroon, waste is transferred using crusher trucks and forklifts loading bins (Attabi, 1999).

2.3.2.3 Waste Disposal Methods

Multiple waste disposal methods are used in developing countries. These methods range from open dumps to sanitary landfills. The open dump approach is the most primitive level of landfilling development and the predominant waste disposal option in most countries of the developing world. It is characterised by indiscriminate disposal of waste and limited measures to control the environmental effects of landfills (Johannessen & Boyer, 1999; UNEP, 2009). In Uganda for example, open dumping in oceans, riverbanks, wetlands and drainage paths is common practice as it is the situation in most cities in other developing countries (NAPE, 2008). Environmental experts state that in India about 350 human carcasses, 1,500 tons of wood and almost that much weight of animal carcasses are dumped on land, and water courses daily. About 1,400 Indians die daily of water related illnesses (WHO, 2009), with an estimated 1,000 children dying of diarrhoea every day. Open burning and burying of waste are also commonly used in Africa, Asia and Latin America.

Most research works on waste management in African cities conclude that open dumps remain the most widespread technique of waste disposal and contribute significantly to pollution in developing nations (Dedehouanou, 1998; Onibokun,1999; ENCAPAFRICA, 2004; William, 2006; Johannessen, 1999; Johannessen and Boyer, 1999) . These authors, however, observed that some countries are upgrading their landfills to sanitary types with assistance from the World Bank. Sanitary landfills of good standards are found in some cities, including Jubaid in Saudi Arabia, Belo Horizonte in Brazil, Mexico City and Buenos Aires in Latin America and Johannesburg in South Africa. Sanitary landfilling is not common practice in most African cities. The sanitary landfill approach is engaged in reducing nuisance as odours, dust, vermin and birds. The main techniques of sanitary landfilling include compaction and daily soil cover to reduce nuisances.

2.3.2.4 Municipal solid waste reduction strategies

Waste reduction can help to reduce waste disposal problems. It entails both waste prevention and waste minimization through various waste recovery practices and activities. Practices and activities like reuse and recycling of waste material are often carried out by individuals and groups for economic and cultural reasons, but they can also be officially encouraged. The product of waste minimization efforts is a reduction in the amount and cost of waste needing transportation and disposal. The result is a longer life span for landfills and less incineration.

Behmanesh (2010) in a study of Pune, India underlines the importance of waste source recovery. Pune, a rapid growing city of about 5 million inhabitants, produces 900-1,100 tonnes of solid waste

31

per day. Organic waste amounts to 65 percent. It is estimated that the largest part of the total solid waste, about 40 percent, is derived from households. The rest comes from hotels, restaurants, shops and markets. Therefore, private separation and processing of household and market waste has the potential to contribute considerably to relieving the burden on the government's disposal infrastructure. In Pune, the Municipal Solid Waste management and Handling Rules 2000 (MSW, 2000), made the disposal of organic waste via composting on residential premises mandatory for housing societies (collective residences) built after the year 2002 (Kroll, 2007 in Behmanesh, 2010). With these regulations in place, the inhabitants of societies now seek acceptable alternative solutions for decomposing and reusing their organic waste.

Medina (2000, 2001, 2003, 2008), Zerbock (2003) and Buhner (2012) studied waste management in third world cities and scavenger cooperatives in Asia and Latin America. They noted that, source recovery could be carried out at different stages of the waste management stream. Source recovery begins when households, institutions and businesses that generate waste separate the materials at the source. Itinerant buyers in turn, may then purchase these source-separated recyclables from individuals or the waste collection crew working for the public or private waste agency, which sorts recyclables prior to the disposal of refuse.

Scavenging is a source of livelihood for the informally employed in low income countries of Latin America and Asia. For example, in Bogota, 12,000 families with an estimated membership of 30,000 to 40,000 are involved in this form of waste recovery, as are 15,000 families in Mexico City (Medina, 2000). Scavenging can be conducted by individuals, micro enterprises, or cooperatives. Scavengers retrieve materials from communal storage sites, streets or public places, open dumpsites, composting plants, and landfills. The activities of scavengers include recovery and recycling of various materials, including paper, cardboard, glass, metal and plastics.

Scavenging is accepted and highly encouraged by some formal, large-scale waste recycling programmes in Latin America. International NGOs, religious groups and local authorities have supported the development of waste scavenger cooperatives with an environmental focus in Brazil, Colombia, and in Mexico, e.g., in Belo Horizonte, Porte Alegre, and Mexico City (Medina, 2000). In Medellin, Columbia, a scavenger cooperative created in 1983 had 1,000 scavengers by 1992 (Pacheco, 1992), 60 percent of them women. It recycled about 5,000 tonnes of recyclables a month. The cooperative also had a contract with Guaine City to collect, transport and dispose of waste. By so doing, the scavengers' cooperative earned an additional income of about US$30,000 per month, while the city realized a cost savings of US $5,000.

In Africa, scavenging exists, but it is less common and less advanced than in Latin America and Asia. One prominent example is the Zabeleen garbage collectors of Cairo--an example of a livelihood improvement strategy for poverty alleviation (Medina, 2000; Fahmi, 2005). About 30,000 Zabeleen waste collectors collect garbage using donkey-pulled carts. They scavenge at household and business premises waste storage sites, as well as at waste collection grounds, dumpsites and landfills. They separate out recyclables and use the organic waste for feeding pigs, the meat of which is then sold to big tourist facilities. They also sell the secondary materials, such as paper, tins, rags, glass and plastics to intermediaries, who in turn sell to craftspeople and industries.

2.3.2.5 Composting

There is growing understanding that composting or local reuse is an environmentally attractive way to manage waste, especially in low and middle income countries (Mbuligwe *et al.*, 2002), where organic waste is the largest fraction of the waste stream and could be removed and composted, offering the possibility to reduce the total waste volume by as much as half. This can have significant consequences for the transportation of waste, the largest cost factor in urban waste management. Cofie (2010) suggested that since composting is both a waste source recovery technique and waste minimization technique, as the compost can be used as fertilizer, it should serve as an attractive alternative for municipal authorities in view of councils' insufficient financial technical and institutional capacities to collect, transport, and safely treat and dispose of municipal waste (Cofie, 2010).

Composting is not, however, without its problems. According to Furedy (2002), composting of organic waste was widely used by the government in cities in Asia, Africa, Latin America and the Middle East in the 1970s and 1980s; however, most of the plants for large scale composting failed, and those that survived did not operate at an optimal level. The failure of plants in cities like Bangkok, Hanoi, Shanghai and Delhi was attributed to the fact that plants were too capital intensive and their output saturated the market, driving down the demand for the product. Replacing old parts or acquiring new technology proved too expensive, and as a result, many cities found it more cost effective not to operate the composting facilities at all.

According to Mensah and Larbi cited in Buhner (2012), all public composting efforts in Ghana have failed. This is also true of Ibadan in Nigeria (Onibokun, 1999). Closure of compost plants have been attributed to low demand due to lack of awareness of compost's soil enriching properties and absence of public support. The costs of composting plants are considered excessive and unjustifiable. The alternative to large composting plants is vermiculture (USAID, 2000). This refers to small earthworms composting farms, operated by 5-6 people. Buhner 2012 argues that this method has proven to be more successful than traditional composting and benefit from better quality control and the perception that worm excrement is derived from 'clean' vegetable waste, whereas compost is gotten from garbage. This method will mean shifting composting burden as well as benefits from council control to the households.

Small scale composting initiatives may be able to avoid the problems of large plants by using a labour intensive process requiring little capital investment, but there is debate about this. Lardinois and Marchand (2000) and Furedy (2002), after examining evidence from Bangalore, Kathmandu and Manila, argued that small scale composting initiatives do not produce enough compost material to be financially viable and cannot significantly reduce the amount of waste that is disposed at landfills (Maclaren, 2003). However, small decentralised composting facilities have proven to be financially viable in cities such as Dhukha, Bangladesh and Vientiane. This latter option seems feasible for low-income cities like Bamenda, which have a high level of unemployment and persons involved in urban and peri-urban agriculture and a high proportion of organic waste (Achankeng, 2005; Akum, 2006).

2.3.2.6 Waste management Cost

In low-income countries, solid waste management is the single largest budget item for many cities (World Bank, 2012; UN-HABITAT, 2010). Waste management costs entail the expenses of waste collection, transportation, treatment and disposal. The cost varies by country, management technique and the priority allocated to waste management. In most developing countries, waste collection alone drains up 80 to 90 percent of municipal waste management budgets, leaving little money for proper disposal. Yet, waste collection lags behind was generation with unpleasant consequences. The scenario is much different in most countries of the developed world, where waste collection has attained a virtually 100 percent rate and collection cost seems to be declining with the introduction of automated techniques. In Japan, for example, collection account for four percent, treatment 45 percent, disposal six percent and transportation and others 45 percent of costs; whereas in developing nations such as Malaysia, collection and transportation costs alone account for over 70 percent of costs (IBRD and World Bank, 1999).

The location of city waste disposal sites affects the costs of waste transportation and management. Generally, city inhabitants resent and protest against the location of waste disposal sites in their vicinities. This resentment is expressed in environmental slogans like NIMBY (Not In My Backyard), LULU (Locally Unacceptable Land Use) and NIMTO (Not In My Term Of Office). These slogans exemplify local resistance to waste disposal facilities. In an attempt to respect the wishes of the population and guard against possible health problems, waste disposal sites are located and relocated to far-off distances from the city. The outcome is increases in transportation costs and tipping fees (Themelis, 2002). The practice of locating MSW disposal sites far off from the city is however contrary to the United Kingdom and European Union (EU) influenced policy of proximity, which emphasizes placing waste management facilities as near to the place of generation as possible (Malcom, 2001).

2.3.2.7 Waste Disposal Impact on Health and Environment

Solid waste generation types, treatment and disposal techniques have far-reaching consequences on health and the environment. Srinivas (2012) and UNEP (1996) reporting on health impacts of solid waste posits that poorly managed household waste can lead to the spread of infectious diseases as it attracts flies, rodents and other creatures that that can serve as disease vectors. In line with a UNEP (1996) report, some waste related infections are: Skin and blood infections resulting from direct contact with waste, and from infected wounds; Eye and respiratory infections resulting from exposure to infected dust, especially during landfill operations; Different diseases that results from the bites of animals feeding on the waste and intestinal infections that are transmitted by flies feeding on the waste. In addition are chronic respiratory diseases and cancers often acquired by persons working at incineration sites or exposure to dust and hazardous compounds.

Current waste discourses stress on the role of waste plastic as a health and environmental hazard. Coloured plastics are highly discouraged because the pigment contains heavy metals considered very toxic. Srinivas identify some harmful metals in plastic and puts a case for their ban. These metals are copper, lead, chromium, cobalt, selenium, and cadmium. Literature reveals that in most industrialized countries, and increasingly in some developing countries, coloured plastics have been legally banned. In India, the Government of Himachal Pradesh has banned the use of plastics and so has Ladakh district. This call resonates the ban on plastics in Kigali in Rwanda and is presently the case in Cameroon as per the joint arête of the Ministry of Environment and Nature Protection and the Ministry of Commerce banning the importation, production and use of plastics (MINEPDED / MINCOMMERCE, 2012).

2.4 MSW Generation and Management in Cameroon

2.4.1 Municipal Solid Waste Generation in Cameroon

Estimating waste generation rates in Cameroon is problematic, as very little comprehensive research on urban waste management is available. In-depth studies published in journals and postgraduate works have focused on major cities as Yaounde and Douala making it difficult for national averages to be estimated. Waste generation data are nonetheless available for some cities as presented in Table 2.3.

Table 2.3: Waste Generation in some selected cities in Cameroon

City	Region	Per Capita per day/kg	Daily City Total /Tonnes
Douala	Littoral	0.88	2,200
Yaounde	Centre	0.85	1,700
Bafoussam	West	0.57	210
Bamenda	North West	0.47	140
Garoua	North	0.37	190

Sources: *Achankeng (2005); Ngnikam & Tanawa (2006); Sotamenou (2010)*

The city of Douala is ranked first in per capita production of solid waste followed by Yaounde. These cities are respectively the economic and administrative headquarters of Cameroon. They have high population, high volume of economic activity, and a higher proportion of prosperous citizens—along with many poor people of course. These are all factors that promote waste production in great quantities and varieties. Jocelyne Adouyou Mouliom, in a Cameroon Tribune newspaper article, "L'invasion du plastique" (2012), notes that Yaounde and Douala have been invaded by plastic waste material, which is loosely disposed of by the population and poorly managed by waste agents. According to the Ministry of the Environment, Nature Protection and Sustainable Development, the city of Yaounde alone generates about 4,000 tonnes of plastic waste daily. The Ministry notes that plastic takes 450 to 500 years to decompose and constitutes about 80 percent of non-biodegradable waste material. This concern prompted the joint ministerial decree of the Ministers of Commerce and Environment, Sustainable Development and Nature Protection. This decree placed a ban on the importation, production and use of plastics (MINEPDED / MINCOMMERCE).

Available literature suggests a relationship between household size and waste amounts. It is thought that single person households produce more waste per capita than families, and ready-made food clearly produces more packaging waste than home-prepared food (Afon, 2007). Braathen adds, however, that traditional preparation generates more organic waste at the household level. The same author also notes that increased incomes make people purchase durable goods such as refrigerators, TVs and other furniture which when discarded in the end produce bulky waste. Construction and demolition waste also add to the municipal waste stream (Braathan 2004). Cities in Cameroon, like those in most of Africa, are the magnets for youths as part of a rural exodus. The resulting youthful urban population often lives in single person households and depends very much on take-away foods, which contributes to non-organic waste. On the other hand, families in Cameroon cities are still very much inclined towards preparing traditional meals (plantain, vegetables, tubers and fruits) from raw produce, which also tend to produce much organic waste.

2.4.2 Municipal solid waste management in Cameroon

Waste material composition, like waste quantity, influence citizen waste handling practices, including storage, reuse, recycling, treatment, collection and disposal. Achankeng (2005) ranked the general composition of waste material in Cameroon by types as follows: organic, ash, paper, plastics, glass, metal, textiles, leather, bones and feathers. The author, like others, added that, although the organic fraction is declining in favour of non-organic waste due to increases in income and the adoption of western lifestyles, the organic proportion remains dominant at more than 60 percent (Ngnikam, 2000: Achankeng, 2005: Sotamenou, 2010).

Urban waste management in Cameroon, like that in most developing countries, is characterized by inefficiency and ineffectiveness, which leads to environmental degradation. Fombe (2006), for example, examined solid waste dumping and collection in the city of Douala. He concluded that the facilities and equipment made available to the city inhabitants by city council authorities and the formal private waste collection company, HYSACAM (Hygiene et Salubrité Cameroun) are inadequate given the proliferation of garbage in the metropolis that results from industrialization, rapid urbanization, and inadequacies in road infrastructure. He called on local authorities to integrate modern techniques with local dumping and collection methods to attain a safe and sustainable city. Essomba (2006), who also assessed solid waste disposal technology and management in Douala, noted that cholera has been endemic there since 1974 and attributed this to poor waste disposal. He noted that densely populated localities with haphazard construction make public access to authorized waste disposal sites difficult. Consequently, waste is deposited in drainage courses leading to air and water pollution.

Parrot, Sotamenou & Dia (2009) investigated the state of municipal solid waste management in the capital city, Yaounde, including livelihood potentials, and offered suggestions for improvement. Obstacles to proper waste management were identified and linked to institutional, financial and physical factors. The study found that excessive distance from households to public waste bins and, in some areas, the complete lack of waste collection infrastructure has a major impact on waste collection and disposal. Additional garbage bins were cited as the primary type of infrastructure needed by the population in all quarters, irrespective of their socioeconomic class. The authors also recommended constructing more waste transfer stations and installing additional bins to reduce distances between households and garbage bins. By so doing, the officially sanctioned private waste collection company HYSACAM (Hygiene et Salubrité Cameroun) could expand its range of services and significantly improve waste collection rates. The study also suggested that more research be carried out on the quality and safety of municipal waste composting for soil fertility to determine whether it can be used to enhance urban and peri-urban agriculture. Most of the stakeholders interviewed, including the municipal authorities, the official waste collection company and households, acknowledged the need for better monitoring and regulation of municipal waste management.

Achankeng (2005) focused research on sustainability in MSWM in Bamenda and Yaounde. The study found that four important, interrelated parts of a MSWM system need to be addressed, with equal emphasis on policy, planning, strategy and practice. These include the primary level (waste generation and storage), the secondary level (waste collection and transportation), waste minimization activities and final disposal. So far, emphasis has rested on waste collection and disposal techniques. Municipal waste strategies generally seem to ignore waste generation and handling at the primary stage, which is crucial to a MSWM system that seeks to improve waste management performance and avoid environmental and health problems at the source of generation. This study builds on this approach by directing attention to the source of waste generation, looking at household and market waste sources. It goes beyond Achankeng, however, by examining the gender issues embedded therein.

The joint ministerial decree of the Ministers of Commerce and Environment, Sustainable Development and Nature Protection (MINEPDED/MINCOMMERCE, 2012) placed a ban on the importation, production and use of plastics. The strategy put in place has begun with the withdrawal of plastic packaging bags from the markets, banning the importation of plastic bags and promoting the use of locally produced packaging made from biodegradable raw materials, including fibre, bamboo, palms and cartons. Cameroon Tribune (2012) suggests that about 40 percent of plastic waste can be valorised (have potential for resource recovery) through reuse or recycling, 30 percent incinerated under environmentally sound conditions and 30percent replaced by biodegradable alternatives.

In addition to the above-mentioned decree are national and local policies that define waste management responsibilities and procedures in Cameroon. The Ministry of Territorial Administration and Decentralisation, under whose jurisdiction councils are found, is charged with the responsibility of ensuring good sanitation and waste management among other duties (Decree No. 98/147 of 17 July 1998). The Ministry of Urban Affairs according to Decree No. 98/153 of 24 July, Articles 22-25 assigns the Ministry with the responsibilities of general cleanliness and drainage, solid waste management, hygiene and sanitation of the cities. Specifically, article 24 assigns the Ministry with the responsibilities of elaborating plans for evacuation and treatment of solid waste, carrying out research on improving collection and transportation; supervision and coordination of collection and transportation; and sensitising the public on the practice of pre-collection of MSW. Actual municipal solid waste management in Cameroon cities including waste collection, transportation and disposal, is the statutory responsibility of the council (Law No. 96/12, 1996).

2.5 Municipal Solid Waste Generation and Management in Bamenda

2.5.1 Major Sources of Municipal Solid Waste Generation in Bamenda

In Bamenda, most waste stems from households, the operation of open-air markets, small-scale urban farming and animal rearing, small businesses and stalls, schools, offices and hospitals and other economic activities. The latter category includes, agro-forestry processing, such as saw-milling, craft shops, light cottage industries, such as woodworking, auto repair, corn and rice processing, brick works, soap processing, and brass and aluminium smelting. Households generate about 90 percent of the total municipal solid waste (Achankeng, 2005). Of the remaining 10 percent, one could expect that a high proportion of it will be waste from open-air markets given the absence of prominent industries.

2.5.1.1 Household Waste Generation and Management

Information about who generates waste, what types predominate, what is done with the waste, and who transports and disposes of it is very pertinent for waste management agencies. Factors influencing household waste generation and variation in quantities and qualities include household size, household income, the age structure of the household, cultural patterns, education, consumer behaviour and life style, packaging of products purchased, available sites for refuse collection, frequency of waste collection services and the price of refuse collection.

Precise information about the quantities and composition of solid waste generated in different areas of African cities is deficient. This handicap has hindered the development of effective solid waste management policies and strategies (Achankeng, 2005; Afon, 2007). Afon (2007) in a study of solid waste generation in a typical African city, Ogbomoso, Nigeria, studied households in three ecological zones – the core, the transitional and the suburban areas. The study thus addressed the limitations of past studies that consider the city as a single entity and ignore internal variation (Onibokun 1999; Dedehouanou, 1998; Onibokun and Kumuyi 1999; Attahi, 1999). Afon's research on Ogobomoso revealed that the quantity and volume of solid waste generated did not increase with income or educational status. Instead, the findings indicated that residents in the core zone with lower average incomes and lower educational status generated more waste per person than those of the other two zones, who had relatively higher incomes and educational status. This finding is contrary to the general assumption that affluence promotes waste generation (Botkin, 2000; Marthandan, 2007; UNEP, 2009).

This present study will deal with the research gap in waste quantities and composition this gap in past research as it investigates the gender component in waste generation and management within the household. In particular, answers will be sought to questions as: who generates household waste? What is done with the waste after generation? What methods of waste handling techniques are used at the level of the household? Who carries out the waste handling activities at the level of the household? Who transports the waste to the disposal site or to the waste collection point for onward collection by the council waste collection team?

2.5.1.2 Market Waste Generation and Management

Probably the second largest waste generation sites in Bamenda are markets. Markets are commercial spaces and resultant waste is thus classified as commercial waste. Commercial waste quantities are linked to the size of business, level of business (wholesale or retail) and the type of material sold (DEFRA, 2011). Market waste generation is also gendered. Njuafac (2012) and Wambo (2013), for example, describe the role of women in solid waste generation, collection and distribution in the Muea and Bamenda Food Markets situated in the South West and North West Regions of Cameroon respectively. These markets serve buyers from different regions of

Cameroon and even persons from other countries in the Central Africa, namely Gabon and Equatorial Guinea.

The researchers set out to identify the role of women in market activities and the characteristics of waste generated in the markets. The markets were partitioned into wards to determine what items were sold, the sex of the person who sold them and how market vendors disposed of their waste in each area. The results revealed that women dominated trading in these markets and were even the more involved in selling perishable items that generate most market waste. It was observed that no waste separation was done before disposal; rather the waste was dumped directly into waste bins made available by the council or into unauthorized locations such as gutters and waterways. Such disposal practices have implications for both waste management and environmental deterioration. Unfortunately, these studies were undergraduate projects, which provide only a cursory overview. The present research will carry out an in-depth analysis and include more markets to provide more comprehensive results and recommendations.

2.5.2 Municipal waste management

Available data and literature about waste management in Bamenda as is the case for most of Cameroon provide an inadequate basis for developing proper waste management strategies and systems. Chuba (1978; cited in Achankeng, 2005) made an early attempt to develop a scientifically based proposal for solid waste management in Bamenda. In a report to the governor of the North West Province, he described municipal solid waste and attempted a calculation of waste generation by the Bamenda population, which was then estimated at 42,720 persons in 9,493 households. Using the average load weight carried by the waste tipper van, which he estimated carried 3.5 tonnes and made four trips a day, he estimated that the waste collected totalled 14 tonnes. He stated, however that this figure was only about one-third of the town's waste because not all waste was generally collected. Consequently, he tripled the estimated daily generation rate of 42 tonnes to accommodate the uncollected garbage. While this study provides vital historical background information for research on waste in Bamenda, its findings are no longer dependable, as the number of households and the population have grown greatly since then. The findings of this study if updated will provide the much-needed information for proper planning of waste management strategies and techniques.

Anschutz *et al.* (1995) also studied SWM management in Bamenda. Their study was based on observation, interviews and questionnaires about waste collection, disposal and associated environmental problems. They described solid waste composition from a study of 340 litres of waste collected from two dumpsites. Waste quantities from the sites were sorted and analysed with the following results: metals/ tins 1.5kg; glass 1.8kg; organic 86.6kg; paper and cardboard 2kg; and plastics/rubber/textile 8.2kg. The weight/volume ratio was calculated as $350kg/m^3$. Using the waste van capacity, monthly waste collection by the council was estimated at 120 tonnes a month, or four tones a day. The study revealed that while waste management practices have not witnessed great changes from the past, the economic crisis that plagued the country during the 1980s and 1990s impinged on the waste management sector seriously, as funding for waste management services dwindled. This was evidenced by a drop in the frequency of waste collection, which led to mounds of litter on roadsides and in water courses.

2.6 Gender Issues in MSW Generation and Management

2.6.1 Basic gender concepts

Gender Women and men are of course different biologically, but each culture interprets and elaborates these innate biological differences into a set of social expectations about what behaviours and activities are appropriate for each gender and what rights, resources and power each should possess. Gender thus involves the socially constructed roles and socially learned behaviours and expectations associated with males and females (Barker, 1999; World Bank, 2001; Pearson, 2006).

Gender roles subject men and women to different environmental risks and impacts and call for different behaviours both in general and in the realm of waste management in particular. These socially constructed expectations about appropriate behaviours and rights affect men's and women's contribution to waste generation and their ability to participate in the management of municipal waste in ways that ensure a sustainable environment.

A **gender analysis** entails analysing data by gender for gender relevant information about the population concerned. It also guides researchers to examine the multiple ways in which women and men, as social actors, engage in strategies to transform existing roles, relationships, and processes in their own interest and in the interest of others (Pearson, 2006; European Commission, 1998). A gender analysis of municipal solid waste management will reveal the impact of the existing waste management system on men, women and children and the impact of gender on the system and its outcomes.

Gender needs/interests are necessities, concerns and values attributed to persons because of their sex, age, class, etc. Practical Gender Needs if met improve the condition of women and men in the short term. They generally include remediating inadequate living conditions, such as providing clean water, shelter, income, health care and other welfare services, including waste collection and disposal services. Meeting PGNs implies that the lives of women and/or men would be improved without necessarily challenging women's subordinate position. Meeting PGNs is a response to an immediate perceived necessity, such as clearing away illegal dumpsites. Strategic Gender Needs (SGNs), on the other hand, are long term changes needed to improve women's position in society (Moser, 1993; Visvanathan *et al.*, 1997; March *et al.*, 1999; Taylor, 2001). Making waste recovery at the level of the household an income generating activity, for example, will financially empower women and will tend to affect household gender power relations to the benefit of women.

2.6.2 Gender and Environment

Gender, society, and environment have been viewed from many standpoints. One of the most relevant for this research is ecofeminism. Advocates of this view posit that men and women have different relations with the environment, and many believe that women have a closer relation with nature and desire to nurture it and even more skilled at doing so. In patriarchal societies, men, on the other hand, are thought to see nature, like women, as resources to be exploited (Mies, 1993; Shiva, 1993; 2005; Littig, 2001).

Echoing the concerns of Dankel *et al.* (1998) who writes on women and the environment in Third World Countries, Care (2004) argues that there is a relationship between women, poverty and the environment in Nairobo (Kenya). This researcher advances a case for reclaiming rights and resources for women. Tiondi (2000) in a research on Sub-Saharan Africa and Latin America establishes a link between women, environment and development. Vandava Shiva carries this opinion even further in studies on India portraying women and children as victims of environmental destruction. Vandava Shiva (1985 & 2005) an environmental feminist argues that the impoverishment of the environment affects children and women immensely who are most involved in fetching water, fuel wood and food for families. Consequently, there is reason to revise cultures and traditions that affect women's access and control to nature's resource. This argument is articulated by Markham & Fonjong (2012) in the paper that suggests a rethink of existing strategies for promoting women's land rights in Cameroon. This paper proposes that there is need to build a gender capacity for male actors. Such capacity will enable men to recognize the role and importance of women to have both access and control rights to land to which they have a close attachment.

Multiple reasons are advanced for women's closeness to the natural environment, especially in developing countries. These include women's concern for their unborn babies, the role of women as the primary managers of the domestic sphere, and the responsibilities they have for maintaining family stability and growing food. It is however important to note that while women's closeness to nature and their ability to nurture it (protection and conservation) is widely acknowledged, it is not necessarily biologically based. Patriarchy and male dominance inhibit the full development of women's potential and relegate them to the roles that bring them in close contact with nature. On the other hand, patriarchy is one of the main causes of women's limited access to and control over resources especially natural resources of land, water and forest in the developing world (Lunde, 1995; Fonjong, 2008; Fonjong, 2010; Markham & Fonjong, 2012; Ngassa, 2013).

The concept of ecofeminism is key in theorizing about and researching waste generation and management activities, which are age-old activities that are generally strongly gendered, albeit not in the same way in every society. They are influenced by the gender division of labour and shaped by ideological, historical, religious, ethnic, economic and cultural determinants (Moser, 1993). Hence, women and men tend to contribute differentially to waste generation and management. Gender issues in waste management can be viewed in several different ways as explained below.

2.6.3 Gender-based perceptions of waste

The definition of waste and discarded materials may be influenced by the gender of the person making the judgements about what to throw away. For example, wastewater and dung from homes, markets and slaughterhouses are viewed by small-scale female urban farmers as compost, manure or soil enhancers, while men may consider it simply as dirt that should be discarded. In addition, waste materials are a resource for certain families, as they serve as a source of sustenance and livelihood. For example, waxy milk containers may be used as fuel, leftover food may be fed to pigs, organic waste may be used as manure for the garden, and discarded cardboard may serve as wall construction material for houses. Consequently, one can expect men and women to perceive and value waste materials differently and focus on their usefulness for different purposes, such as domestic use, saving on household expenditures, earning money, or other purposes. These gendered definitions of 'waste' and 'resources' must be reflected in any discussion of waste management, including community consultation processes used to develop waste management strategies (Chi, 2003; Thu, 2005).

2.6.4 Gender Roles and Gender Division of Labour in the Waste Economy

The 'waste economy' includes activities and operations that involve waste management as a source of livelihood. Many roles and occupations in this economy are gendered: that is, most remunerative activities carried out along the waste management chain are done primarily by either women or men. According to Muller & Schienberg, (1998; 2003), women are often engaged in waste related activities that require low levels of education and skill, such as sweeping, waste picking from dump sites, sorting and washing, and waste collection, cleaning canals and maintenance of parks and other public places (Hardoy *et al.*, 2001). Men, on another hand, are more frequently in waste activities that involve the use of machines, including vehicular transportation of waste (Thu, 2005).

2.6.5 Gender in Access to Waste Resources for Livelihood

The socio-cultural and economic status of men and women (gender role, level of education, financial power, etc.) in the household and community influence the value they attach to waste and determine their access to and control of potential waste resources. According to Thu (2005), the generally subordinated status of women may limit their general access to and control of important resources, such as timber and minerals of the soil and water, and their access to well-paid urban jobs. Consequently, deriving income from waste related activities may be among the few economic options available to them, and any innovations in managing waste materials made by the public or private waste operators that fail to consider these female activities may create or destroy some of their sources of livelihood. Gupta (1998), for example, conducted a study on excreta collection in Ghazaiabad, which illustrated how new waste management policies could negatively affect women. The study found out that before Ho Chi Minh City (Vietnam) decided to integrate the informal waste management sector into the formal sector, women had formed the majority of workers collecting human excreta in the informal sector. The municipality then decided, in the interest of 'better' overall urban waste management, to integrate informal sector services into the formal sector through direct employment of waste workers or subcontracting to small enterprises for this task. However, when the municipal department decided to place the workers who collected excreta on municipal payroll, 70 percent of these employees turned out to be men.

Muller and Schienberg (1998) in same line of thought as Gupta (1998) believe that similar mechanisms may be in operation when small enterprises obtain municipal subcontracts in the waste sector. In such cases, competition for employment in these enterprises may intensify, as they offer greater stability of income, forcing women out for reasons linked to their biology or gaps in education and technical skills, or even outright discrimination. Initiatives that give an economic value to waste material or activities at the primary level of management (generation and storage), such as payment for waste sorting, picking and collection of waste, will go a long way to address women's strategic needs. Whereas schemes that fail to attach value to primary waste activities, especially at the level of households, tend to destroy livelihood sources for women.

2.6.6 Gender and Waste Impact on Health

The existence of gendered waste work and the gendered division of labour in households and markets, as well as in waste management institutions, implies that women and men are subject to different impacts of waste on their health. Exposure to waste often has a negative impact on the health of waste workers. A health survey by the Hanoi Urban Environment Company, for example, revealed that street cleaning workers' vulnerability to infections from aerobic bacteria during work hours is 3.1 times greater than for women who do not do this work. Findings from the survey also revealed that the presence of streptococcus bacteria in throat swabs was 1.78 times higher in areas where waste work is carried out than in those without. Women are those most involved in street cleaning and, therefore, the ones most affected by ailments resulting from such activities. Additional findings revealed that, for those waste workers who lift garbage into trucks, exposure to aerobic bacteria during work is 5.7 times higher than when not at work, while streptococcus bacteria exposure is 2.2 times higher. In this case, men were most affected. Finally, Dat's (1995) study of the health impacts of waste collection on women in Hanoi found that the incidence of respiratory problems in men and women is different: 20.5 percent and 29.7 percent of male and female workers respectively had symptoms of respiratory problems. Women tend to be very sensitive to diseases caused by the working environment and frequent morbidity tends to reduce their labour hours and, correspondingly, their income.

2.6.7 Women as Central Actors in the Generation and Management of Household Waste

Women's triple roles - productive, reproductive and community involvement (Moser, 1993) in developing societies tend to make them primarily responsible for domestic waste generation and management. Within the household and community, women perform productive roles in the waste management sector that earn income and reduce expenditures. For instance, they carry out wage labour in urban agriculture, which generates organic waste, street sweeping and scavenging. They also care for members of the family who fall ill even from poor waste disposal effects. Women carry out domestic chores of cooking and cleaning (reproductive roles). They are also more involved in activities such as catering and education and clean up campaigns (community roles).

2.6.8 Limitations on women's effective participation in the waste economy

Thu (2005), like Dat (1995), argues that women's participation in the waste economy as a source of livelihood is restricted by several factors. Women's low level of education limits them to relatively lower paid activities like sweeping. More lucrative scavenging at dump sites situated far away from settlements eliminates most women, as they are often preoccupied with home chores or exposed to harassment at such worksites. Financial obstacles, such as discrimination in obtaining credit, can block them from purchasing waste resource recovery equipment as diggers, shovels and rakes and limit their full participation in waste activities outside the domestic environment, even though these activities generate the most income. The tools and equipment used by women are often very rudimentary. They include bamboo carrying frames and poles, whereas most men who work in scrap iron buying have bicycles, and some have motor bikes. These modes of transport available primarily to men facilitate movement and increase their area of waste operation. The women travel shorter distances and their area of operation is limited because of lack of motorized means (Dat, 1995; Chi, 2003).

2.6.9 Men and women's participation in decision making in the waste management process

Decision making about waste management can also be strongly influenced by gender. The outcomes of the deliberations of bodies that make decisions about waste policies, strategies and techniques in cities depend on how forums for input and negotiation are structured. For instance, the time and settings of stakeholders' consultation meetings may define the meeting as either a men's or a women's space. If such meetings become dominated by one gender or are held in localities in which women or men are not comfortable or do not feel free to express their opinions, then one gender, frequently women, may participate little. The UN-HABITAT (2013) handbook on gender and local governance suggests that such forums influence the provision of municipal services including electricity, water and waste management.

Such environments also tend to affect the nature of input and decisions about in waste management. For example, a study of a program concerned with marketing waste and employment in waste activities in Vietnam (Thu, 2005), reveals that women were most active in neighbourhood and street-level development decision-making committees, which were closest to their household management roles. Women's participation at higher levels, such as the community or city level meetings, or as leaders was much less frequent. This implies that women's expressed priorities, voiced in lower level meetings, could become lost in the decision-making processes, and the final results may reflect women's concerns only partially, if at all. It is in such circumstances that gender mainstreaming through affirmative action and quota representation ensures men and women's presence and voice in the development process (Squire, 2007).

2.7 Theoretical Background

This study is informed and guided by theories of Consumerism (Thorstein Veblen, 1899), Treadmill of Production (Gould, Pellow & Schnaiberg, (2003) and Ecofeminism (McGuire & McGuire, 1991). Other important concepts and frameworks have also been used to appreciate and analyse findings. These include the Waste Management Hierarchy Concept, the Gender Analysis Framework, the Gender Roles Framework, and the Social Relations Approach Concept with a bearing on gender institutional policy assessment.

2.7.1 Consumerism Theory

In developed societies, and increasingly in developing ones as well, consumption of large volumes of consumer goods has become a way of life, and people's worth is often measured by what they consume. Economists even equate personal happiness with what people purchase and consume. Consumer society began to emerge in the late seventeenth and eighteen centuries and intensified during the 19th and 20th centuries. The consumer society's traits are linked with industrial growth. These economic changes led to mass production and a dramatic increase in the availability of consumer goods, the advent of department stores and low cost of commodities, which motivated mass consumption in Europe and the United States. Products became available in vast quantities at low prices that were affordable by most citizens in the industrialized west.

Thorstein Veblen (1899) in the *Theory of the Leisure Class* (Britannica, 2008) described consumption in the late 19th Century as "conspicuous". And his insights continue to apply even in today's societies. He identified a seemingly irrational form of economic behaviour, characterized by visible, wasteful consumption designed to prove one's wealth and gain prestige. Mies and Shiva (1993) draw on this theory by pointing out that consumers often seek to imitate those who are above them. The current 21st century growth and expansion of new world economies like China and other industrialized emerging nations of South America and South East Asia has extended consumerism to these new markets and made more and more goods available at low prices. Moreover, even poorer societies, such as those in Sub-Saharan Africa, have not been immune to the influence of consumerism.

Consumerism has been harshly criticized, especially in developed nations, by groups that advocate alternative lifestyles, such as simple living, eco-conscious shopping, buying and consumption of local products. Moderate anti-consumerism advocates, on the other hand, do not oppose consumption in itself, but argue against increasing consumption of resources beyond what is environmentally sustainable. They point out that consumers are often unaware of the negative impacts of producing, transporting, and landfilling the many modern goods they consume. They add that extensive advertising only serves to reinforce increasing consumption (Brand, 1997; Rumbo, 2002). Protagonists of consumerism, including Libertarians, however, argue that anti-consumerism will lead to elitism, and that consumer culture is good for the economy, employment and the soul. They argue that claims of damage to the environment are greatly exaggerated. They see anti-consumerism as a precursor to central planning or a totalitarian society, which will lead to economic crisis. The economist Victor Lebow (1995) even advocates for consumption to be adopted as a lifestyle and a ritual.

Expounding the argument for consumerism, many people in developing countries see anti-consumerism as illegitimate. They argue that anti-consumerism consists of high consumers who have made a mess of the planet telling emerging countries that they should now limit consumption and exploitation especially of forest resources to save the planet. Whatever, the overall merits of consumerism, it has become a major force in developing countries where it promotes increased waste generation, whose management remains problematic. The capitalist/consumerist way of life in modern industrial societies with a spill-effect on developing countries is linked with high usage of nature and strong emphasis on growth and consumption (Reusswig, 1994). The fear is that, growing consumerism in developing countries will lead to even greater environmental damage and an unsustainable world. Of particular relevance here is the possibility that, in the developing countries where waste management lags behind generation, irresponsible waste practices will further compound the damage from consumerism on the environment.

Examining consumerism from a feminist perspective, Littig (2001) argues that, while pollution and environmental destruction were initially mainly associated with industrial mass production, there is equally a strong relationship between the consumption of products and environmental destruction. Littig like Dally and Farley (2003) posits an inherent link between consumer driven consumption and planet-wide ecological degradation resulting from resource use, production, and use of the planet as a waste sink. While the relationships between environmental consciousness and socio-demographic characteristics (age, education, income and gender) are often murky in existing research it is known that the consequences of pollution and environmental deterioration that stem from consumerism fall disproportionately on men and women. The latter are often the victims of ecological damage especially in developing countries due to their closeness to nature as advanced by ecofeminism.

The theory of consumerism helps us to understand waste generation as a consequence of consumption. Municipal waste generation is influenced, amongst other factors, by the types of products made available to consumers by vendors and the consumption aspirations and pattern of households. As long as consumerism becomes part of people's consciousness, they can be expected to try to acquire more and more "stylish" goods, even at the expense of the environment. This has obvious implications for increased waste generation. Furthermore, if consumerism suppresses environmentally sound attitudes, consumers may be less interested in using products until they are worn out or purchasing used items, with obvious impacts on waste generation and management practices. This study therefore sets out to determine if households are environmentally conscious – (knowledgeable about the impact of waste on the environment); and practice environmentally sound behaviour – (through waste management practices)

2.7.2 Treadmill of Production Theory

The treadmill of production theory was developed by Allan Schnaiberg. It was based primarily on his analysis of the US economy. However, its results are believed to be applicable in other areas. Schnaiberg (2003) argues that the public's and government's desire for economic expansion will generally prevail over ecological concerns. State policy will generally prioritize immediate economic growth over environmental degradation and will deal effectively only with environmental problems that threaten public health or economic disaster. He predicted that the outcome of negligence of the environment would be severe environmental degradation.

The Treadmill of Production Theory focuses on conflict and cooperation between three economic actors, the state, monopoly capital, and labour, as it relates to economic expansion and environmental concerns. In the conflict, the state and labour, but usually not monopoly capital, may at times view the acceleration of economic growth with the associated resource use and pollution as a potential danger to the environment. However all three actors usually end up joining to favour economic growth because it provides jobs and social security, and tax revenue.

Schnaiberg believes that the benefits of this coalition supporting accelerated economic growth will be short-lived, as the environmental damage that results will have negative long-term effects on both state funding and workers' livelihood. In the event of such undesirable reality, the alliance between the state, capital and labour may eventually break down, and the state may act to control the worst of the environmental damage. He suggested that it is therefore important that the state and labour movements are educated about the long run environmental and livelihood dangers of supporting monopoly capital. He hopes that a deceleration of the treadmill could lead to environmental improvement and a break in the pattern of ever-increasing acceleration supported by the treadmill alliance. Justified as these hopes may be, realising them remains a huge challenge. This is because we base our whole economy on the assumption of continued technological progress and economic growth. We know no other way than growing industrial production to keep people employed. Hence, the prediction of the theory that consumption will continue to rise because of how the treadmill works remains valid.

The primary implication of the treadmill of production theory for this study is that with accelerated economic growth and production, affluence and consequently high consumption waste generation will be on the rise. This is because consumers reach forever-increasing levels of affluence and government supports economic growth and indirectly consumption because it creates employment. There will also be pressure from the business sector against any measures that would slow down sales and reduce profit. Wealthier neighbourhoods will produce more waste in quantity and variety. There is also likely to be the generation of dangerous and hard to process waste associated with affluence. This study seeks to determine in what ways: households serve as production centres for municipal solid waste; markets contribute to municipal solid waste generation and type; and the effects of municipal solid waste on the environment.

2.7.3 Ecofeminism

The term ecofeminism was coined by Francoise D'Eaubonne as a criticism of the ideologies of the ecology and feminist movements (McGuire & McGuire, 1991). She criticized feminist movements for ignoring environmental issues and faulted ecological movements for omitting a feminist analysis of environmental issues. D'Eaubonne began by describing the epic violence inflicted on women and nature because of patriarchy. The theory describes society as a hierarchical structure in which man holds power over the woman and dominates her and nature (McGuire and McGuire, 2003). Ecofeminism thus establishes a relationship between destruction of the environment (ecology) and subordination and exploitation of women. It champions action against patriarchy's domination of women and nature as the only solution.

Some ecofeminists also advance an argument for women's special connectedness to nature. For example, Mellor (1997) believes that women have a closer relationship than men to nature owing to their living and working situations as caregivers and nurturers. This means that women are much more threatened by the consequences of ecological destruction through pollution and other environmental dangers (mental and physical disabilities, diseases, premature deaths, etc. than men are. Mies and Shiva (1993) attribute this closeness to women's ability to bear children and bring forth life.

Ecofeminists while generally endorsing the principles, on which the theory is based, sometimes disagree about the causes of women's subordination, their close connectedness to nature and environmental destruction. Based on the main points of disagreement, Mellor (1997) classifies ecofeminists' perspectives into two: ecofeminists with a cultural orientation and ecofeminists with a constructivist orientation. Ecofeminists with a cultural orientation stress male domination (patriarchy) and even maleness in itself as the cause of ecological destruction and socially oppressive behaviour. They also ascribe an elementary closeness of women to nature. Ecofeminists with a constructivist's approach, including socialist feminists, view the division of power and particularly of labour between men and women in society as the key to unsustainable patterns of development, environmental degradation and the exploitation of women. They see women's connectedness to nature not as 'natural', but rather as a cultural and historically changeable product. Liberal feminist with a constructivists orientation, in particular, challenge the argument that women are closer to nature. This is because such a position undermines the feminist arguments against the naturalization of gender differences that have been used to legitimize women's subordination and advocates for the improvement of women's participation in opportunities in the political and economic systems. On the other hand, feminists with tendencies toward deep ecology (inherent worth for living beings) fear that the active involvement of women in the socioeconomic system will have a negative outcome. Involvement implies women's participation in economic, political and social activities, which in the course of raising their status will contribute to the destruction of humanity and the environment.

In any event, almost all ecofeminists believe that the liberation of women and protection of the environment can only be achieved through a radical change in the prevailing patriarchal, eco-social and political system (Knapp, 1998). Nevertheless, the question remains: How can the prevailing patriarchal system based on gender power relations be changed to address the women's practical and strategic gender needs without women's involvement in important economic and political activities? This study of gender and municipal waste management with emphases on the feminization of household waste generation and management suggest possibilities. As such, the study sets out to investigate whether women by their close connectedness to nature: are more conscious (aware) of waste related environmental problems and the resulting threats to health and well-being than men; exhibit more sound attitudes to the environment than men in their perception of waste and waste management behaviour in relation to waste storage and disposal; are more exposed to waste related environmental risks than men. The study will also determine whether patriarchy keeps women out of participation in key decisions about waste management.

2.8 Gender Analysis Framework: The Harvard Analytical Framework

As a gender related study, we use a gender framework to direct the study on gender issues in household and market municipal solid waste generation and management. The Harvard Analytical Framework is chosen for this study as it makes visible men and women's work. The Harvard Analytical framework is also called the Gender Roles Framework or the Gender Analysis Framework. This framework is one of the early gender analysis and planning frameworks and was developed by the Harvard Institute for International Development in collaboration with the Women in Development (WID) Office of USAID. It is based on the WID efficiency approach, which advocates for the provision of welfare services for women. By this development approach, women are portrayed as beneficiaries of development programmes and not as actors. The Harvard framework was originally outlined in Overholt, Anderson, Cloud and Austin, *Gender Roles in Development Projects: a Case Book*, published in 1984

The Harvard Analytical framework provides a matrix for collecting data at the micro (community and household) level. It has four interrelated components: The activity profile, the access and control profile, the analysis of influencing factors and the project cycle analysis. This research will employ the first three components of the framework. The fourth, which is the project cycle analysis, is ignored, as it is not appropriate. It is better suited for actual development projects due implementation.

The activity profile addresses gender roles and gender division of labour by providing answers to the question, "who does what?" including gender, age, time spent and location of activity. In this study, the activity profile will include those activities carried out at the level of the household and market, which generate or manage waste. Some of these activities include, food preparation, gardening and sweeping. Gender division of labour within the household ascribes different roles to women, men and children. There is equally sex division in types of businesses in operation in the market and consequently in the type and volume of waste generated. The access and control profile component is in this study operationalised in variables as nearness to council waste collection infrastructure (bins, and mobile vans), other waste disposal sites, participation of households and market vendors in decisions concerning council waste plans. Thirdly is the analysis of influencing factors component. This component identifies factors other than the gender politics within the household and markets that influence municipal waste management. These factors will serve as the extraneous variables for this study and include institutions charged with municipal waste management, culture and politics.

2.9 Conceptual Framework

According to Kumar (2005), a conceptual framework forms the basis of the research problem. The conceptual framework for this study as presented in Figure 2 proposes a gender integrated MSWM plan for most African cities.

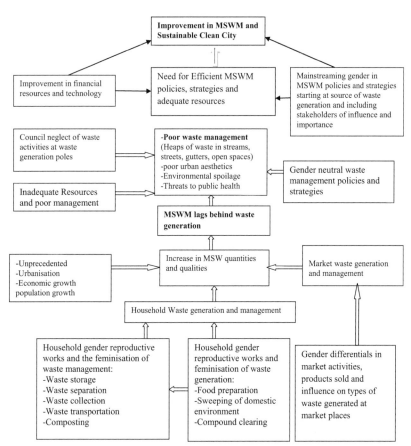

Figure 2.1: A Gender Integrated MSWM Framework for African Cities

(Source: Akum, 2013)

Chapter Three

Methodology

In this chapter the scope of the research, geographic area of study, research design, sampling techniques, instruments used for data collection and data analysis procedures are presented.

3.1 Research Scope, Type and Context

This thesis uses both quantitative and qualitative approaches. It looks mainly at the extent to which gender roles and division of labour in the household and market sites influence women, men and children contribute differentially to municipal waste generation and management activities. Due to the fact that the research questions and objectives of this study necessitate data from diverse sources, triangulation is employed in the data collection and data analysis procedures. Triangulation is a research approach employing more than one method of data collection and analysis, and using results from different sources simultaneously to check one another and converge on an understanding of the research questions (Sarantakos, 1998).

In this study, the triangulation approach is used in data collection with the use of secondary data sources and primary sources (questionnaires, interviews and on-site measurements). It is also employed in data processing, analysis, and presentation of results by combining qualitative and quantitative techniques. This approach is applied to investigate the different activities along the waste generation and management chain and examine relevant behaviours and, gender issues and gender gaps, which may be obvious or latent in the current council waste management policies and strategies. This multiple methods approach or research technique has the advantage that internal and external validity is ensured. This technique is very important in this study, as data sources in Cameroon can be highly inconsistent owing to illiteracy, lack of value attached to research by the population, lack of confidence in researchers' promise of confidentiality and fear of the unknown, among others.

3.2 Area of Study

3.2.1 Physical Environment

This study was conducted in the area governed by the Bamenda City Council (BCC). Bamenda is located in the North West Region of Cameroon (Fig. 3.1) at latitudes $5^0 89" 74"N$ and longitudes $10^0 10" 21"E$ in a hilly area on the Bamileke-Bamenda mountain range. Divided by a tall and steep escarpment, the topography of Bamenda presents two facets. One facet is the higher lying and sloping topography of the Bamendankwe area and its environs, which contains the Bamenda I Administrative Sub-Division. The other is the depression below the escarpment, which contains the Bamenda II and Bamenda III Administrative Sub-Divisions. Bamenda I is situated at the edge of an escarpment overlooking Bamenda II and Bamenda III (Figure 3.2).

Bamenda's climate is characterised by two seasons; a rainy season (mid-March to mid-November) and a dry season from mid-November to mid-March). Average annual rainfall is 2.288mm, and average annual temperature is 19.70^0C (Bamenda UP-Station Meteorological Centre; cited in Akum, 2006). The relatively high rainfall and temperature have important implications for effective waste handling, disposal and decomposition. For instance, organic waste soon liquefies and odours appear very quickly.

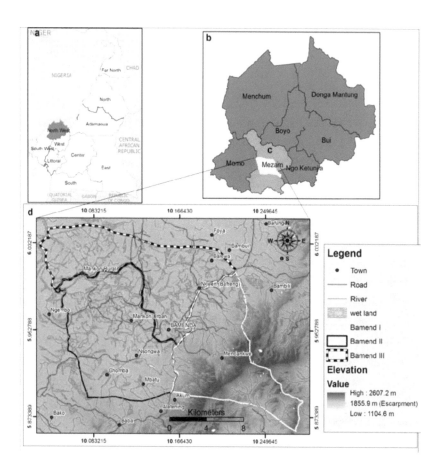

Map 3.1: Location and Physical Presentation of Bamenda City Council: a -Cameroon; b –North West Region; c- Mezam Division; d – Bamenda City Council Area (*Source, Akum, 2013*)

The hydrography of Bamenda includes rivers, streams and springs within the city area and a generally high water table. The most prominent river is the Mezam; it and its tributaries flow through the city. The major soil type is ferruginous soil derived from the basement complex and old sedimentary rocks. The land is covered by weak clay subsoil due to erosion and translocation of clay from the top soil. This physical milieu, typified by a non-uniform landscape, heavy rainfall, and ferruginous soils tend to promote high runoff and leaching. Runoff from rainwater and leaching tend to carry contaminants from poorly disposed waste into the soil, underground water and surface watercourses, including streams and rivers used by the population for domestic purposes and urban agriculture.

3.2.2 Social and Political Environment

Bamenda is the administrative capital of both the North West Region of Cameroon and the Mezam Division. It is the primate city of the North West Region in spatial, economic and population growth terms. It is also the birthplace of the two leading political parties in Cameroon, the Cameroon Peoples Democratic Party (CPDM) and the Social Democratic Front (SDF). Due to its key administrative and economic position, Bamenda has experienced rapid growth, which fosters rapidly expanding waste generation. According to the 1987 census (DSCN 1987), with 146,021 inhabitants Bamenda ranked fifth in population after Douala (Cameroon's economic capital), Yaounde (the administrative capital), Garoua and Maroua (Fombe & Balgah, 2011). The 2005 population census raised Bamenda municipality to the third position after Douala and Yaounde, with a population size of 322,889 inhabitants. Greater detail indicates that 269,530 of these inhabitants live within the borders of the city's three administrative subdivisions.

The population is 48 percent male and 52 percent female, including 158,302 males and 164,587 females. This sex distribution contradicts the traditional assumption that urban population is often male dominated because more men than women migrate to urban settings (Waugh, 1995). Government administrative jobs, and urban agriculture and retail businesses, which dominate the economic sector, attract women who have little or no access to large farms and lack the capital for big businesses. Projecting an annual growth rate of 4.9 percent (BUCREP, 2010), the population for Bamenda can be estimated to be 496.931 in 2012 (MINDU, 2011) and is expected to reach 433,636 by 2020. Urbanisation is known to be a principal cause of increases in municipal solid waste generation (UNEP, 2009). In the case of Bamenda, soaring rates of urbanization, both population increases and spatial expansion, engender settlements on marginal and inaccessible lands and are likely to encourage waste dumping

64

3.2.3 Administrative Division

The Bamenda City Council, the hub of the North West Region (MINDU, 2011), has jurisdiction over the three administrative sub-divisions, Bamenda I, Bamenda II and Bamenda III (Figure 3.2). These subdivisions are under the jurisdiction of sub-divisional officers appointed by higher government authorities. Along the lines of these subdivisions are councils (communes) manned by mayors and councillors who are elected. Created in 2008, these subdivisions correspond to the three main *Fondoms* that earlier made up Bamenda town namely, Bamendakwe, Mankon and Bafreng (Nkwen) respectively (Neba, 1999) by virtue of their central location. All three subdivisions both urban and rural areas (Chomba, Mbatu, Pinying, Santa, Akum and Awing among others) make up Bamenda City Council. This study is limited to the urbanised areas within the Bamenda City Council. Settlements in these areas are distinguished by relatively higher population densities, modern infrastructure, and public services, such as pipe borne water, paved roads and unpaved but accessible roads, economic activities dominated by non-agricultural nature as business, daily markets, schools, etc. The places (quarters) considered as urbanized (Mendankwe Urban) in the Bamenda I Sub-Division and used in this study cover the Government Residential Area (GRA), and Mile One and its environs comprising Alanting, Ntamaanfe and Ntaasia (Figure, 3.2). Bamenda I plays host to most of the government offices.

Bamenda III (Nkwen) is separated from Bamenda II by a small stream called the Lipacan that flows through a narrow valley from the Bamenda escarpment (Akum, 2006; Neba, 1999). It is a fast growing area experiencing urban sprawl as its farm lands are taken over for commercial and residential land uses. Present at Nkwen are travel bus stops as the Nkwen Mile Two Park, Nkwen Mile Four Park and the Amour-Mezam travel agency. The most populated neighbourhoods found in the two urban health districts of Nkwen Urban and Nkwen Baptist include Bayelle, Ndamukong, St Paul and Foncha Street; and Nkwen New Layout, Mugheb and Sisia respectively. Nkwen is the gateway to the recently created Bamenda University, which is located at Bambili, Tubah Subdivision. The Nkwen Mile Two and Nkwen Mile Four Markets are located in

65

Bamenda III. Nkwen also hosts a private university referred to as the Bamenda University of Science and Technology (BUST). Also situated in Nkwen are the National Polytechnic Bambui and the St Louis University Institute of Health Sciences. Spatial expansion, coupled with a fast growing population attracted by these institutions, encourage increasing waste generation. This very rapid growth tends to outrun the city's ability to provide adequate waste management infrastructure in the area.

Figure 2.2: Layout of Bamenda Municipality

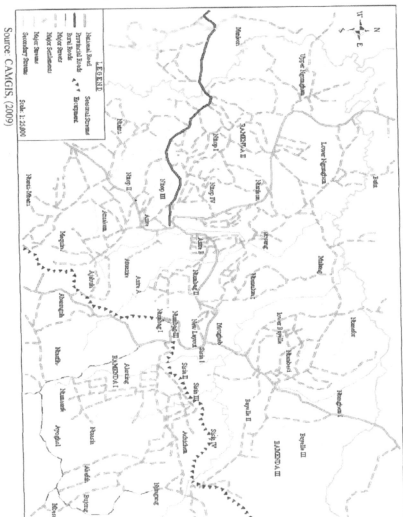

3.2.4 Education

As the centre in the North West Region, Bamenda holds a prime position in the English sub-system of education in Cameroon. It is host to many private, religious and government-run primary and secondary institutions. Its regular annual position at the top of the results of the General Certificate of Education Board (GCEB) results, both in the number of candidates and grades obtained is evidence of its leading role. Bamenda also hosts many private institutions of higher learning (see Section 3.2.2.1). These educational institutions serve as a major pull factor for students nationwide, increasing population growth and ultimately waste generation.

3.2.5 The Economy

The structure of the Bamenda economy includes the primary, secondary and tertiary sectors of production and both formal and informal sectors. Primary sector industries are mainly sand and stone quarrying, the exploitation of timber from eucalyptus trees, and farming (both crop and livestock). In addition, much gardening for vegetables (huckleberry greens, tomatoes and waterleaf) takes place in the dry season in the low-lying riverside areas. Poultry is raised for table birds and eggs, and pigs for meat. Organic waste from poultry and piggery is an economic asset, as it is highly demanded as manure for gardening (MINDU, 2011). At the other extreme, toxic waste from other sources such as auto-repair shops dumped in waterways tends to pollute the water, which is often used, especially in the dry season, to irrigate farms. Such pollutants can cause health disorders for consumers of cultivated products.

Bamenda lacks major manufacturing plants, Instead, secondary sector economic activity is mostly informal and scattered through different neighbourhoods. These activities revolve around small scale processing of cassava into garri and "water-fufu", transformation of milk into yoghurt and foodstuffs into animal feed, small wood and metal works, embroidery, carving and crafts (MINDU, 2011). These processing activities affect the volume and composition of municipal solid waste generated. Their waste is disposed of in same sites as domestic waste. Handling such waste, especially, where separation is not practiced, makes management very difficult.

Activities of the tertiary sector are very prominent in formal and informal sector activities. The formal sector activities, which are very visible, include financial and commercial services. Popularly known banking institutions include BICEC, SGBC, Union and Amity banks. In addition to these financial mainstream-banking institutions are micro-credit establishments as CCC (Community Credit Company, Azire Credit Union, Ntarikon Credit Union, etc.). Money-Transfer agencies have in recent times been on the rise; notably Express Union and Money Exchange. Commercial activities on their part include wholesale and retailing services. There are nine markets within the city that operate on a daily basis as presented and discussed in chapter six. Other major service providers are transport agencies of public and private ownership. In spite of these economic activities, the income levels of most of the inhabitants remain low and youth unemployment as is the situation in other cities in Cameroon remains high. The consequence is high crime wave and a generally low standard of living. Developing the waste economy by adding an economic value to waste can offer employment opportunities, promote the economy and reduce criminality. The presentation of land use in Bamenda City Council is shown in Map 3.

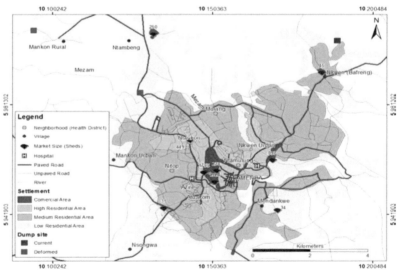

Map 3.2: Land Use in the Bamenda City Council

(Source: Akum, 2013)

3.3 Basis for Selection of City Case and Study Design

Before January 2008, Douala and Yaounde, with populations of 1.6 and 1.4 million respectively, were the only urban centres with the rank of city in Cameroon. Presidential decree No. 208 of January 2008 upgraded Bamenda, Limbe, Maroua, Kumba and Ebolowa to join the class of cities led by city delegates. This implies, among other things, greater obligation for the city council to maintain a high level of sanitation and funding to comply with new status. These are obligations which efficient waste management can, to a large extent, address, by causing a clean environment, saving public health and cutting down on council waste budget.

Ideally, several of these new cities from different geopolitical and ecological zones would be selected for study. This was not possible due to time and resource constrains. This researcher chooses instead to concentrate on a single city, Bamenda, (the largest in its class) to carry out an in-depth study, the results of which can serve as a baseline for carrying out further studies of cities in this size range. Bamenda is also of special interest because it is the biggest city in the predominantly Anglophone region and the capital of the North West Region and the Division of Mezam. While some research about waste management has been conducted in Yaounde and Douala, Bamenda is relatively a virgin territory for research. Also, although most of the other larger Cameroonian cities have privatized the collection and management of city waste collection, Bamenda remains one of the very few that still assumes almost complete responsibility for collecting, transporting and disposing municipal solid waste. This study allows us to look at the extent to which this system functions. It is important to note that not all findings gotten from a restricted waste management system that entertains very little private intervention like that Bamenda can therefore be generalized.

3.4 Household Waste Generation and Management

3.4.1 The Study Design
The most important part of the study, collection of data from households, uses a cross-sectional survey design to investigate a

sample of Bamenda households' in terms of their knowledge about waste management, their waste management practices regarding municipal solid waste generation and management. It also examines gender factors in household solid waste generation. The cross-sectional survey approach provides for relatively speedy data collection and the ability to generalize to a population from a sample. It also saves time and resources (Oso & Onen, 2008; Kumar 2005; Sarantakos, 1998).

The sampling design for the household study employed multistage cluster sampling. Multistage sampling involves selecting progressively smaller geographic clusters of households until the individual households in the sample are selected. Bamenda City Council, the study area, was divided into subdivisions, then to urban clusters (Health District Areas) and then to neighbourhoods popularly referred to as quarters, and eventually to households. This method of selection approximates a random sample if the size of the sample in each subdivision is proportional to its percentage of the population. Specifically, the sample was drawn by using the three subdivisions as the first stage; then the health districts (neighbourhoods) within the urban areas of subdivisions as the second stage; and finally the quarters from the urban health districts in subdivisions as the third stage. Households were then selected from within the quarters and finally adult individuals in the households to represent their households.

3.4.2 Population, Sample Size and Sampling Procedure

3.4.2.1 Population

Households in Bamenda I, Bamenda II and Bamenda III constitute the population of this study. The target population of the survey are residents of Bamenda city (including Bamenda I, Bamenda II and Bamenda III) residing in a dwelling be it an independent house, an apartment, a flat, or a shanty. Institutional dwellings, such as hostels, hospitals, clinics, nursing homes, jails, barracks or orphanages and homeless people were excluded. A residence attached to an office/business was considered a household. The 2011 household census data developed for mosquito-net distribution by the Bamenda Health District (BHD, 2012) Department survey, served as the guide for sample selection for this study. The BHD (2012) survey identified nine health districts in the Bamenda City urban zone namely, Mendankwe (GRA & Environs) Alakuma, Atuakom, Azire, Mulang, Ntambag, Ntamulung, Nkwen Urban and Nkwen Baptist. These health district areas were not intended to be homogeneous in infrastructure, population size or function. Their boundaries are determined by physical elements such as roads, streams/river or a remarkable infrastructure feature among other indicators. Based on the BHD 2012 survey records, there were 47,585 households in these nine urban health districts.

3.4.2.2 Sample Unit and Sample Size Calculation

The sampling unit for the household survey is the household. Sample size was estimated using sample calculation for one proportion with the support of Epi Info 6.04d (CDC, 2001) as explained by Nana (2012).

$$n = \frac{NZ^2 P(1-P)}{d^2(N-1) + Z^2 P(1-P)}$$

Where N=total population, Z= Z value corresponding to the confidence level, d= absolute precision, P=expected proportion in the population, n effective=n*design effect.

In the context of this study, the following parameters were used to estimate the sample size:

Size of the population 47585

Desired precision 5%

Expected prevalence 80%

Confident level 95%

Design effect (DEFF) 1.5%

N.B. The DEFF was considered higher than 1 because the study did not use a simple random sampling though a good geographical coverage was ensured.

A critical proportion level for adequate waste disposal of 80% was assumed by:

Firstly, P is situated within 95% CI using the formula below.

$$P-(Z_{\alpha/2})\sqrt{pq/n} < P < P+ (Z_{\alpha/2})\sqrt{pq/n}$$

Where:

- P= prevalence
- n= sample size at a given expected prevalence (here 80%). Considering this proportion, a confidence level of 95%,
- q= 1-P
- $Z_{\alpha/2}$ =level of significance = 1.96.

The conjectured proportion within 95% Confidence Interval will then be obtained.

79.64 <80 <80.36

Secondly, we calculate the sample size for the ranged values of P at 95% CI

Using the proportion range and applying the formula above, the estimated sample sizes with the lower and upper bound at 95% CI is as follows:

For a total study population of 47,585, the estimated sample size is: 362 < 367 < 372.

This sample size was distributed to sub-divisions, health district areas and eventually quarters and weighted by their respective populations.

3.4.2.3 Sampling

The nine urban Health District Areas (HDA) as identified by the Bamenda Health District (BHD) survey were further sub-divided into 47 quarters. It should be noted that the names of some health district areas double as names of quarters, such as Azire and Alakuma and Mulang. The name of the quarter in which a health unit (centre) is seated is used to name the health district area. The number of households selected in each health district area is a function of their proportion in the total number of households in the city. This is calculated by multiplying the sub-population percent in column D, by the desired sample size (367). The number of households were then determined per quarter by calculating a sample size proportional to each quarter's proportion in the HDA based on the District Health

Survey statistics (C). The result is in column E as presented in Table 3.1.

Table 3.1: Household Sample Size Distribution by Neighbourhoods per Subdivision

Sub-Division	Health District Area	No. of quarters in DHA	No. of households in DHA	Sub-pop percentage	Sample size (n)
	A	B	C	D	E
Bda I	Mendankwe	4	1,867	3.92	14
Bda II	Alakuma	5	2,398	5.04	18
	Atuakom	4	4,899	10.30	38
	Azire	10	10,714	22.52	83
	Mulang	3	5,313	11.17	41
	Ntambag	4	2,926	6.15	23
	Ntamulung	10	3,039	6.39	23
Bda III	Nkwen Urban	4	11,177	23.49	86
	Nkwen Baptist	3	5,252	11.04	41
Total		47	47,585	100.00	367
Formulae				x/47585x100	D/100x360

Adapted from BHD (2012) Survey

In each quarter, selection of households was purposive rather than random. Houses were selected with infrastructure assured to be reflecting different socioeconomic statuses, proximity to the main street or in the interior and location along watercourses to ensure good geographical and social class coverage. In each household an adult member, preferably the father or mother and in some cases grandparents served as respondents. In households with only one adult, then that one served as the respondent. In households with unrelated adults, the one available and willing to respond was selected. Households without children were also considered for the survey. The sample size was sex disproportionate with a 60 percent female and 40% male. This difference in favour of the female gender is explained by the fact that women in most African communities are more involved in household activities and spend more time at home than men. In such regard, it was assumed that they will be more informed about household waste issues and will be in a better position to provide the required information for this study. Given that a 60% to 40% female and male representation respectively was desired for this survey, research assistants made a purposive selection of the respondent by his/her sex. That is a potential respondent in a household could be rejected or a household omitted because of the sex of the available respondent. This discrimination was done when the required number of respondents for one sex was attained. In addition to the household data collected through questionnaire (Appendix I), more results on household waste characteristics were gotten from household on-site observation and measurement exercise (Appendix II) carried out in four quarters of different income classes. This supplementary survey from different income class quarters complemented data from questionnaire and ensured validity.

3.4.3 Data Collection Instruments

Data for this study was collected from both primary and secondary sources and in different phases. Multiple visits were made to the field to gather information that guided waste collection procedure. To ensure validity and reliability, a pre-test of the main data instrument, questionnaire for household survey was carried out in Tiko in Fako Division away from study area in the month of August 2012. Ten motivated research assistants in collaboration with the lead researcher were involved in the data collection exercise. Data collection with field assistants spanned from October 2012 to April 2013 interspersed with breaks. Data verification and completion continued after this period. Instruments for primary data collection included questionnaires, interviews, on-site waste observation and measurements.

3.4.3.1 Questionnaire

A self-designed questionnaire (Appendix I), including both closed ended and open-ended questions, was the main data collection instrument. Questionnaires were preferred since the study is concerned, in part, with variables that cannot be directly observed, such as opinions and perceptions of respondents. In addition, the population size is large, and given the time and other constraints, a questionnaire administered to a sample of residents was the best tool for collecting representative data (Kumar, 2005; Onen & Oso, 2008; Nana, 2002).

Self-administration or guided administration of questionnaires was carried out in residents' homes. Fifty questionnaires were self-administered by those who were able and expressed willingness to complete them. For this category of respondents, it took a minimum of two hours and sometimes two to three days, and even a week of repeated visits to retrieve answered questionnaires. In some instances, these questionnaires were never returned or poorly completed as the case of the Atuakom HDA. Here, multiple unsuccessful trips by the research assistant to retrieve questionnaires ended up in resignation. To resolve this deficiency, a purposive selection of Abangoh quarter in the HDA of Atuakom for the observation and on-site household waste measurement procedure was made by the researcher.

For most respondents, who were unable or unwilling to self-administer, guided administration of questionnaire was carried out by the researcher and research assistants. This meant that the research assistant read out questions to the respondent and then filled out the questionnaire based on the responses given. The pidgin language, based on English, which is widely spoken in the area, was often used to explain information to some respondents. The guided technique of questionnaire administration gave the researcher and assistants the opportunity to explain the purpose, relevance and importance of the study and clarify any questions respondents had. This technique also ensured a high return rate of distributed questionnaires.

Concerned that not all questionnaires distributed would be returned, slightly more than the desired sample size, 400 questionnaires were distributed, and 372 obtained. Most of the non-returned questionnaires were from self-administered respondents. Out of the 372 retrieved, 339 were considered usable, giving a 92% response rate based on the intended sample simple size of 367 households, which was considered adequate for the study. Some questionnaires were rejected based on missing vital identification or responses. Questionnaires with a high degree of incompleteness, internal inconsistencies and illegible handwriting were rejected.

As earlier mentioned, the questionnaire included both closed and open-ended questions. The closed-ended questions called for responses about the household's physical and socio-economic characteristics, household waste generation activities and actors, household waste management practices and activities by gender, assessment of council waste management strategies and possible gender differences in them, opinions about mainstreaming gender in municipal waste management policies and strategies, and background information about the respondent. Particularly, a gender activity profile was developed for some close-ended questions that solicited information about household gender division of labour and level of participation in activities that generate waste. The open-ended questions asked about the respondents' opinions and perceptions about solid waste, the current waste management strategies by the council, and proposals for improvement.

3.4.3.2 Household On-site Observation and Measurements

A household on-site waste measurement procedure (Appendix II) was developed to determine household waste quantities and composition for selected households. This exercise also provided the opportunity to assess people's ability and willingness to separate waste if asked to do so. Waste from 108 randomly selected households from four different income class quarters as defined by the Regional Delegation of the Ministry Urban Development and Housing (MINDU, 2011) was collected and measured: high income (Foncha Street), middle income (Ndamukong), low income (Ntambag) and peripheral (peri-urban) zone (Abangoh-Atuakom).

The selection of quarters for the household waste measurement procedure was done by assigning the numbers to all the quarters. One quarter was drawn at random from each income class. For the peripheral zone, a purposive selection of Abangoh quarter in the HDA of Atuakom was made by the researcher to compensate for a low questionnaire response return rate in the HDA of Atuakom as earlier indicated (see Section 3.4.3.1).

Table 3.2: Sample distribution of households by income class quarters

Quarter/neighbourhood	Income class	Number of households	Sample size
Foncha Street	High	3,692	38
Ndamukong	Middle	2,600	27
Ntambag (Old Town)	Low	2,926	23
Abangoh/ Atuakom	Peripheral	1,515	20
Total		**10,733**	**108**

Adapted from BHD Survey (2012)

The objective of this procedure was to determine the amount of waste generated per capita per household in a week, and the composition of the household waste stream. The number of households studied in each quarter was determined by its proportion to the population of the health district area within which it is situated. Three visits were made to each household within a week. The first was to brief the households on the objectives of the project and provide disposable waste storage bags. Two bags of different colours were given per household, a green bag for biodegradable waste and a black bag for non-biodegradable waste. The second visit, made on the fourth day, was used to analyse the waste through observation, sorting, and weighing. The bag holding inorganic waste was further separated into recyclable, non-recyclable, and each type was weighed using weight measurement scale. During this second visit, used bags were replaced with new ones. This was to avoid inconveniencing the families by requiring them to hold their waste for long periods. The third and last visit was carried out on the eighth day, when the analysis was repeated. Information observed was recorded on a pre-prepared information sheet (Appendix II). The information collected is important and of interest in itself and has been used to corroborate responses from observation, interviews and questionnaires.

3.5 Market Waste Generation and Management

The study on municipal solid waste generation and management in Bamenda also included a component designed to investigate the waste generation and handling techniques among a cross section of vendors in the market. Literature suggests that waste from commercial sources comes second after households in Bamenda (Achankeng, 2005; Anschutz *et al.*, 1995). The objective of this component was to determine the characteristics of market waste, gender related issues to market waste generation and the level of environmental awareness and consciousness of market vendors. The study design sought to identify the waste generation potential and typical types of waste from the different types of articles sold and activities/services carried out by vendors in the market place.

3.5.1 Research Design

This section of the study, which focuses on solid waste generation and waste handling in the markets, also employed a cross-sectional research design. Quota sampling was used to determine the sample size for each market. This was followed by a systematic selection of vendors (stalls and open ground allotments) as respondents for guided administration of structured interview (Appendix II).

3.5.2 Population, Sample Size and Sampling Procedure

3.5.2.1 Population

Nine markets operate daily within the Bamenda City Council (MINDU, 2011; Field Work, 2012). All were included in this study. The vending spaces include stalls in permanent and non-permanent (makeshift) structures. Permanent structures are subdivided into lock-up stalls and open slabs, which are built up and cemented. Non-permanent structures on their part are categorised into thatches/kiosks and open ground spaces, which are, not cemented (Table 3.3). These vending spaces/stores/places comprise the population for this study.

Table 3.3 Distribution of markets and vending allotments by infrastructure

Location of Market	Markets	Infrastructure			
		Permanent		Non-Permanent	
		Lock-Up stalls	Open Slabs	Thatches/ Kiosks	Open-Ground Space
Bda I	Bda Up-Station	30	-	-	43
Bda II	Bda City Council Main Market	702	477	596	149
	Bda City Council Food Market	242	444	89	980
	Bda City Stadium Market	208	-	-	30
	Ntarikon Market	365	-	76	8
	Ntatru	31	-	-	-
	Muwachu Mile 8	250	-	-	-
Bda III	Nkwen Market Mile 2	466	-	396	35
	Nkwen Mile 4 Markets	45	-	-	-
Total		2.339	921	1.157	1.245

Adapted from Market records and market survey, 2013

3.5.2.2 Sample unit, Sample Size and Sampling Procedure

The sample unit is the vending space (stalls, open slabs, thatches/kiosks and ground allotments) as presented in Table 3.4. An open slab is a semi-built up structure with a table-like cemented surface with pillars raised above to support the roof. On the slab, items are exhibited for sale. Based on statistics provided by the administrative authorities of the markets, supplemented by a quick visual count, estimates of the number of vending allotments were established for each market. General observation was made of market infrastructure and the layout of selling grounds in order to partition the market into sections according to the type of infrastructure. That is, the market was divided into areas with permanent infrastructure (lock-up stalls and open-slab stalls) and non-permanent infrastructure (thatched open space, kiosk) and no infrastructure (open ground space). At a confidence interval of 95% with an estimated population of 5663, we chose a sample size of 354 using the same procedure as for households, explained earlier (see Section 3.4.2.2) using the formula $n = \dfrac{NZ^2P(1-P)}{d^2(N-1)+Z^2P(1-P)}$

The first step in data collection was to conduct a quick count of the vending spaces in the market to make an informed guess about the different types of items sold and activities/services provided in the various markets, categorized as food items, non-food items and services. The types of material used for packaging was also noted. This is because most packaging tends to end up as municipal waste, either in or out of the market, providing a clue as to the characteristics of municipal solid waste material. Information observed was recorded on a pre-prepared sheet (Appendix III). The second step was to note the relationship between the gender of the vendor and activities/services carried out in the market (for example, corn milling, repairs, catering, etc.) and items sold. This was to determine if market waste characteristics in terms of quantities and types could be linked to the gender of the generator, or/and the activities involved by the vendor. Such information is relevant to municipal waste operators, urban and development planners as they seek to create awareness of the necessity of proper waste management.

Location of Market	Markets	Infrastructure			
		Permanent		Non-Permanent	
		Lock-Up stalls	Open slabs	Thatched/ Kiosks	Open-Ground space
Bda I	Bda Up-Station	2	-	-	3
Bda II	Bda City Council Main Market	45	30	38	9
	Bda City Council Food Market	15	28	6	62
	Bda City Stadium Market	13	-	-	2
	Ntarikon Market	23	-	5	1
	Ntatru	2	-	-	2
	Muwachu Mile 8	16	-	-	-
Bda III	Nkwen Market Mile 2	30	-	25	-
	Nkwen Mile 4 Markets	3	-	-	-
Total		149	58	74	79

3.5.3 Instruments for Data Collection

3.5.3.1 Observation Information Sheet

The activities/services, items sold, and types of material used were recorded on an information sheet (Appendix III) with the intention of determining their contribution to waste generation in the market. Observation was done of the different waste material types at the communal bins at the market sites and other nearby disposal sites and recorded. Observation was carried out in all the nine markets that operate daily. Saturday was chosen as the main day for observation because it is main market day. It is a day in which urban dwellers and people from rural areas bring farm produce and buy other items from the markets.

3.5.3.2 Structured Interview Guide

A short structured interview (Appendix IV) with both closed-ended and open-ended questions was administered to the vendors in the sample by research assistants. The structured interview was preferred as it is less time consuming for the vendors, who often have unpredictable schedules and limited time. Interviews were conducted on Tuesdays, Wednesdays and Thursdays which are considered from experience as non-peak market days and during non-peak sales periods of the day, i.e., between 11:00AM and 15:00PM. These periods were selected in order to reduce interruptions and allowed the vendors time for responding to questions; however, the procedure might have injected a bias against vendors who appear only on the busiest days. When buyers approached the vendors, the interviewer quickly made way for sales and buy transactions. Additional appointments were made after market periods with some vendors who were very interested in the study or the interviewer judged to be very knowledgeable about the subject. At such opportunities, which were rare, vendors, were given the opportunity to give their impression of the waste management techniques in the market and what could be done to improve on them.

One section of the interview guide solicited basic information, including the types of articles sold, material used for packaging and the sex of the vendor. These were observed by the research assistant and the information recorded. The next part of the interview guide required information from the vendors about other demographic characteristics, other activities and services carried out by vendors during their stay in the market place, packaging material, perception of waste, waste storage and waste handing techniques. Out of 354 structured interviews intended, 345 were successfully administered resulting in a 97.5 percent return rate. This result was considered very good and sufficient for statistical analysis.

3.6 Other Data Collection Techniques

3.6.1 Interviews with Elected and Appointed Officials

In-depth interviews based on an interview schedule using semi-structured and open-ended questions were conducted with key stakeholders in municipal waste management (Appendix V). In all, 15 interviews were conducted (five women and ten men). Interviewed at the level of the Council were the Government Delegate to the Bamenda City Council, the Chief of Service for the Municipal Garbage Department and a driver of the pick-up waste collection team. Regional Delegates for Urban Development (a department of the national government) and Environment, Nature Protection and Sustainable Development were also interviewed. Government officials were selected based on the statutory obligations that their departments have as stakeholders in MSWM. Discussions were also carried out with elected officials (three mayors and three councillors from Bamenda I, II and III). The interviews with appointed government officials and elected authorities were used to assess the operation and efficiency of the current waste management system by examining relevant regulations and laws, technology and infrastructure and stakeholder participation. This assessment looked at the entire waste management process, including collection, transportation, treatment, disposal, recycling and recovery. It also aimed to identify the challenges the City Council faces in managing solid waste.

Three local community administrators of the quarters (generally referred to as quarter-heads) of Musang, Azire and Sisia quarters were equally interviewed. Quarter heads were selected based on the degree of influence as revealed in informal discussions with inhabitants and to represent all three sub-divisions that make up the city. In addition, I interviewed the Managers of four leading markets (Bamenda City Council Main Market, Bamenda City Council Food Market, Ntarikon Market and the Nkwen Market). The market managers were selected based on the size of the markets.

3.6.2 Observation of Household Waste Composition and Council Waste Collection Techniques

In addition to the on-site observation of households and markets, non-participant observation of the waste collection techniques and activities of the council mobile waste collection team was conducted. I accompanied the council waste collection team as it went about the routine duty of picking up waste on roadsides and transporting it to the municipal waste dumpsite. The goal was to determine the different types of equipment used to collect and transport waste, the behaviour and attitudes of the waste collection team, and the response of the population to the service. Observation also helped to determine the gender and age of those involved in waste transfer from households into the mobile van. In addition, observation was used to study the occurrence of environmental problems associated with solid waste management, such as polluted water bodies, roadside littering and waste dumps. Observation sheets were used to record these phenomena (Appendix VI) in the quarters of Sisia and Foncha Street in the Nkwen Urban Health District Area (Bamenda III); Nghomgham and Musang quarters in the Mulang HDA and Commercial Avenue and Old Town in the Ntambag HDA (Bamnenda II). These quarters were purposefully selected bearing in mind their exposure to drainage features, which are likely to serve as waste dumpsites and areas of urban agricultural activities. Information collected was used to corroborate or refute responses in questionnaires.

3.7 Data Processing and Analysis

A triangulation process was followed in analysing the data by combining qualitative and quantitative approaches in the analysis and interpretation of results where necessary. Specifically, a gender analysis profile was developed for gender-related data. Three different statistical packages were used to analyse data: Data coding was done using the Atlasti software version 5.2. Data were entered using Epi-Info 6.04d (CDC, 2001) and analysed using the Statistical Package for Social Sciences (SPSS) Standard Version, Release 17.0 9SPSS Inc. 2008).

3.7.1 Processing and Analysis of Questionnaire-Based Data

3.7.1.1 Review and Labelling of Questionnaire
An initial editing was used to check the content of questionnaire responses for completeness and internal consistency. Questions without responses in incomplete questionnaires that contained enough information to be included were coded as missing data while those with a high degree of incompleteness, internal inconsistencies and illegible handwriting were rejected.

3.7.1.2 Coding Questionnaire Based Data
For the purpose of analysis, closed ended items were dealt with differently than open-ended ones. For open-ended items, a code list was developed that summarized the major concepts that emerged from the responses. A code in this context is an umbrella term that summarizes a set of individual responses with a common theme (Appendix VII - Code quotation grounding report). For instance, the responses "I cannot pay for waste disposal because it is government responsibility," "I will not like to pay for waste disposal because I pay taxes" and "I shall not pay for waste disposal because the council shall take charge" can be summarized under government responsibility / authorities' responsibility as code. Codes are accompanied in data analysis by their brief descriptions (code description) to inform users about the main or key ideas that they include and illustrated by well selected (two or three) quotations (Appendix VII)

3.7.1.3 Questionnaire-Based Data Analysis
Data were analysed using the following techniques as described by Bluman (2007), Kumar (2005), WHO & TDR (2008) and Nana (2012). For categorical variables, such as 'Who decided on the placement (location) of the public bin/ open dump that you use', frequency distributions and cross tabulations are used. Multiple Response Analysis was used for multiple-response question (i.e. questions for which more than one response to a single question). Levels of significance of association between nominal/nominal (e.g. association between sex and sensitization/education on waste

management) and nominal/ordinal pairs of variables (e.g. association between sex and knowledge of amount of waste produced) were assessed using Chi-Square test of independence or equality of proportions. The Somers'd or Spearman's Rho tests were used for ordinal/ordinal pairs (e.g. correlation between distance to the waste disposal site and number of times waste-bin is emptied). Scale variables were screened for normality using Kolmogorov Smirnov and Shapiro Wilk tests. They were notably, in weight and capacity; the 'amount of organic waste generated', 'amount of inorganic waste generated', 'amount of recyclable waste generated', and 'amount of non-recyclable waste generated'. Given that these scale variables did not follow a normal distribution, non-parametric tests notably Kruskal Wallis test was then used to compare their distribution for more than two groups for significant difference. This is exemplified when comparing amount of waste generated by sub-divisions or neighbourhoods. Ratio and interval variables were analysed with the use of measurements of central tendencies. Results are presented using statistical tables, charts and diagrams. All significance tests used the 0.05 level of significance ($\alpha = 0.05$).

3.7.2 Processing and Analysis of Interviews

I began by transcribing the interviews conducted with council and some appointed officials, elected council authorities and market managers from notes. Responses that were not clear or accidentally omitted questions were completed by contacting respondents who were so available to make clarifications and give answers by telephone. In this regard, collaboration from the chief of service in the department of hygiene and sanitation charged with garbage collection at the City Council was very remarkable. Given the relatively small number of interviews, responses from these interviews were analysed by hand.

3.8 Validity and Reliability of Research Instruments

Validity, which is the ability of an instrument to measure what it is designed to measure (Kumar, 2005), was established using logical and statistical evidence. Logical evidence was established through face and content validity. Face validity was determined by posing questions in the research instruments that were in relation to the objectives of the study. Content validity was established for questions with less tangible concepts that solicited respondents' opinions, such as perception. To such concepts, several questions were asked in order to cover different aspects. To ensure validity through the statistical approach, the coefficient of correlations were calculated between the questions posed and the outcome variables. The supervisors who are expert researchers read, corrected and validated instruments as adequate for the study.

Reliability as the degree of accuracy or precision in the measurements made by a research instrument was established. The household questionnaire, the main research instrument was pre-tested in Tiko in the South West Region of Cameroon away from study area. Qualified research assistants with university education with knowledge and experience in field data collection participated. As a group, the lead researcher had a training session with the research assistants during which occasion they were briefed on the objective of the research and other related issues.

Chapter Four

Gender Analysis of Household Solid Waste Generation

This chapter presents research findings related to the first two objectives of the study: 1) to provide a demographic profile of households and household respondents and 2) to examine household gender reproductive roles (care and maintenance of household and household members) and their relationship to household waste generation. The goal is to make visible the central role of women in household waste generation and the need to see women as stakeholders of importance in municipal solid waste generation. Data in this chapter are presented in three sections: 1) a demographic profile of respondents and households; 2) a description of the physical environment of households; and 3) a gender analysis of household solid waste generation.

4.1 Demographic Profile of Households Respondents

Findings in this section address objective one of this study, which focuses on the demographic profile of household respondents and their households in the Bamenda City Council (BCC). Data come from responses to questions in section V-A of household survey questionnaire, which solicited information about household characteristics and the respondents themselves. The demographic variables considered include the age, sex, household income, household size, level of education and occupation of respondents and household members.

4.1.1 Sex and Age Profile of Household Respondents

To ensure an ample representation of women, the sample was selected so that females represent 60 percent (203) of respondents and males 40 percent (136). Table 4.1 presents the sex and age distribution of respondents from households sampled.

Table 4.1: Sample distribution of respondents by age and sex

Age	Sex				
	Male		Female		
	Count		Count		
	(n)	Percent (%)	(n)	Percent (%)	
16 - 20	9	7.3	11	5.9	
21 – 25	16	12.9	35	18.9	
26 – 30	28	22.6	41	22.2	
31 – 35	24	19.4	23	12.4	
36 – 40	14	11.3	20	10.8	
41– 45	18	14.5	20	10.8	
46 – 50	8	6.5	10	5.4	
>50	7	5.6	25	13.5	
Total	**124**	**100**	**185**	**100**	
Missing					
= 30					

$x^2=9.592$; df=7; P=0.213   *Source: Household survey, 2013*

A majority of the respondents for both sexes fall within the age range of 21-45: male (81 percent) and female (75 percent). These are considered persons in the active age group in the labour market and biological reproduction. Many women in this age group have childrearing and household responsibilities of which waste generation is an outcome. A higher percentage of women than men are in the age group of persons above 50 years with six and 14 percent for male and female respectively; however, there is statistically no significant difference in the sex of respondents in the different age groups ($x^2=9.59$; df=7; P=0.21)

4.1.2 Level of Education in Households

Level of education is likely to influence the way people perceive waste and their ability to understand waste sensitization messages. Table 4.2 shows that an overwhelming majority (95 percent) of respondents of both sexes have had at least some formal education. This is, however, to be expected, because of the higher availability of schools and advanced education in urban areas. There is no significant difference between the sexes in education level. Slightly over a quarter of both men and women have attended secondary schools, and well over half of respondents of both sexes have had some form of post-secondary education.

Table 4.2 Sample distribution of respondents by level of education and sex

Level of Education		Sex			
		Male		Female	
			Percent		Percent
		n	(%)	n	(%)
No formal education		2	1.7	5	3.5
Primary		18	15.3	22	15.4
Secondary		34	28.8	37	25.9
Post-secondary		64	54.2	78	54.6
Total		**118**	**100**	**143**	**100**

χ2-test: χ2=0.84; df=3; P=0.84   Source: Household survey, 2013

The level of education of household members with decision-making power in the household can influence their level of environmental consciousness and that of household members, as well as waste generation and management practices. Consequently, the questionnaire included questions about the educational status of "father" and "mother". In most of Africa and Cameroon, the terms parent, mother and father are used to describe decision makers, instruction-givers, and heads of households in the family, not just biological parents. They could be heads and co-heads of households by ability and capability to provide for household needs and make decisions, or simply by seniority in terms of age. The concepts of father and mother may therefore span to include grandparents, aunts, uncles, and even senior brothers and sisters when biological fathers and mothers are absent or do not have absolute decision-making powers. Table 4.3 presents the level of education of father and mother (decision makers) in the household. The question allowed respondents to decide who best fit these roles. Where no response was provided, generally because no such person was present, the response was coded as missing.

Table 4.3: Sample distribution by level of education of father and mother

Level of education	Father		Mother	
	Count (n)	Percent (%)	Count (n)	Percent (%)
No formal education	7	2.8	5	2.4
Primary	40	15.7	44	20.8
Secondary	84	33.1	60	28.3
Post-secondary	123	48.4	103	48.6
Total	**254**	**100.0**	**212**	**100**
Missing	85		127	
Total	339		339	100

$\chi2$-test: $\chi2=1.30$; $df=3$; $P=0.73$ Source: Household survey, 2013

The way this category of persons, fathers and mothers (household decision makers) perceive waste issues may have an impact on the nature of the information and instruction they provide to other household members. The information, instructions and ability to enforce them can influence waste generation and management practices within the household. The results are, not surprisingly, very similar to those reported in Table 4.2, and there is statistically no significant difference in the level of education of fathers and mothers in households.

4.1.3 Household Occupational Profile

The occupation of household decision makers (fathers and mothers) is likely to influence household income, and consequently, choices of food and other items they can purchase, which in turn, are likely to influence waste quantities and composition. Table 4.4 presents findings relating to the occupation of "fathers" and "mothers" in the household. Table 4.4 presents findings relating to the different types of works in which fathers and mothers as household decision makers are engaged in. Thirty-seven percent and 34 percent of "fathers" and "mothers" participate in paid occupations of managerial, administrative and professional nature (white-collar jobs) and 50 percent and 34 percent respectively of skilled and semi-skilled jobs.

Table 4.4: Sample distribution of households by occupation of father and mother

Occupation	Father		Mother	
	Count (n)	Percent (f)	Count (n)	Percent (n)
Agriculture/Farming	16	6.8	23	8.5
White collar jobs (professionals)	87	36.9	96	33.5
Manual works	119	50.4	92	33.8
Housewife	-	-	45	16.5
Others (student, etc.)	14	5.9	17	5.9
Total	**236**	**100.0**	**272**	**100**
Missing	103		67	
Total	339		339	

χ2-test: χ2=1.75; df=3; P=0.62   Source: Household survey, 2013

The white-collar jobs identified include management, administration, teaching, engineering, architecture and research. The skilled and semi-skilled occupations include trading, metal works; woodwork, electricians, textile, printing, butchering, baking, security services and catering. Results reveal that the difference between the distributions of occupations for fathers and mothers falls just short of being significant at the .05 level. In about seven percent and eight percent of households, fathers and mothers respectively are involved in agriculture in the form of crop farming, and animal husbandry. Though agricultural activities engage only these relatively small fractions, they represent a surprising proportion, as agriculture is generally not regarded as an urban activity. The surprising representation of agriculture here could be attributed to the absence of industrial activities and low incomes of many residents.

Findings further show that a greater proportion of fathers are involved in white collar and manual (semi-skilled) work, which can be specified than females (Tables 4.4). Away from the household, results in productive work in the public sphere indicate that the level of involvement of mothers in white-collar (professional) occupations is very high. This result contradicts the assertion by March *et al.* (1999) which argues that women unlike men are often less involved in paid work, and that when involved, they participate or occupy positions of relatively low pay and status. The assertion however holds true when the position of housewife is considered. It is worth noting that the Cameroon civil service law allows no sex discrimination in salaried payment. This means women as men with same qualification in same job position earn equal pay.

4.1.4 Household Income Levels

Income also has an effect on household consumption patterns and, consequently, quite likely on waste generation amounts and types. The household questionnaire contained a question that asked about household monthly earnings. The findings presented in Table 4.5 reveal that an overall majority of households (63 percent) earn less than 100,000 FCFA (about 200 dollars) a month and only 13 percent earn 200,000 FCFA or more.

There is a statistically significant difference in the monthly income levels of households in the three subdivisions of BCC ($\chi 2=33.38$; df=8; P <0.01). There are too few households from Bamenda I in the sample to be confident of the generalisability of the results from this area, but the income difference is striking. Seventy percent of households in Bamenda I report relatively high incomes of 150.000FCFA and above while a majority of those of Bamenda II and III earn less than 100.000FCFA. Less than a fifth (19 percent) and about a third (32 percent) of households in Bamenda II and Bamenda III earn more than 150.000 francs a month. The relatively higher incomes of households in Bamenda I can be attributed to the presence of persons of professional and managerial job positions who are often on regular state pay roll. On the other hand, most of the residents of Bamenda II and Bamenda III are involved primarily in semi-skilled or unskilled occupations that tend to generate relatively lower incomes. These low-income figures generally indicate a community with a low purchasing power.

Table 4.5: Sample distribution by income and by sub-division

Sub-Division	Levels of household income										Total	
	<50000		50000-99999		100000-149999		150000-199999		≥200000			
	n	%	n	%	n	%	n	%	n	%	n	%
Bamenda I	0	.0	1	10.0	2	20.0	3	30.0	4	40.0	10	100.0
Bamenda II	64	30.3	80	37.9	28	13.3	12	5.7	27	12.8	211	100.0
Bamenda III	25	28.7	24	27.6	10	11.5	20	23.0	8	9.2	87	100.0
Total	89	28.9	105	34.1	40	13	35	11.4	39	12.7	308	100.0
Missing= 31												

$\chi 2=33.385$; df=8; P <0.01.   *Source: Household survey, 2013*

4.1.5 Household Size

The mean, mode and standard deviation for household size in the city are 5.4, 5.0 and 5.6 respectively. The mean household size for Bamenda I is 7.1 persons. This number is greater than the 4.9 for Bamenda II and 6.1 for Bamenda III (MINDU, 2011). The results for Bamenda I should be interpreted very cautiously due to the small few of cases; however, the relatively higher mean for Bamenda I may indicate that the higher the household income, the bigger the household size. However, a correlation between household size and income shows no significant relationship as indicated on Table 4.6.

Survey results show that seventy-one percent of households in the City have at least a child below the age of 16. The mean, mode and standard deviation for household with children below the age of 16 in the City are 2.8, 2 and 1.3 respectively. This age group is of special relevance, as children, especially in developing countries, play a very important role in household labour (YRI, 2003). Children below the age of 16 in most homes in the region are charged with the responsibility of transferring waste from domestic environment to public waste collection sites or disposal sites. This activity is considered by the society as not burdensome and within the capabilities of children.

4.1.6 Correlation between Demographic Background Indicators

Table 4.6 shows the relationship and the level of significance between the different household demographic variables of education, household size, age and income. A positive correlation (.70) exists between the level of education of father and that of mother. This relationship is statistically significant ($p < .01$), indicating that educated men tend to marry educated women. Not surprisingly, a significant positive correlation (.64) is seen between the number of persons less than age 16 and the household size. Not surprisingly, positive and significant correlations are also found (.39 and .39 respectively) between household income and level of education of fathers and mothers of households.

Table 4.6: Correlation between demographic background indicators

Indicators	A	B	C	D	E
A		R= 0.70** P<0.01) N=146	R= 0.06 P= 0.39 N= 229	R= -0.10 P= 0.19 N= 178	R= 0.39** P= <0.01 N= 234
B	R= 0.70** P= < 0.01 N=146		R= -0.14 P= 0.05 N= 192	R= -0.01 P= 0.87 N=161	R= 0.39** P= <0.01 N= 200
C	R= -0.06 P=0.39 N= 229	R= -0.14 P= 0.05 N= 192		R= 0.64** P< 0.01 N=227	R= 0.10 P= 0.08 N= 287
D	R= -0.10 P= 0.19 N= 178	R= -0.01 P=0.87 N= 161	R= 0.64** P< 0.01 N= 227		R= 0.07 P= 0.32 N= 224
E	R=0.39** P<0.01 N= 234	R= 0.39** P<0.01 N= 200	R= 0.10 P= 0.08 N= 287	R= 0.07 P= 0.32 N= 224	

Source: Household survey, 2013

Key for Table 4.6.

R= Correlation Coefficient,

**. = **A correlation is significant at the 0.01 level (2-tail)

N = Number sampled

A = Level of education of father (head/co-head of household)

B = Level of education of mother (head/co-head of household)

C = Number of persons currently living in the household

D = Number residents of the household less than 16 years old

E = Household level of income

4.2 Physical Environment: Condition of the Streets and Drainage Features

Data presented in this section provides information about the physical environments of the households. It is based on questions in the household questionnaire. This information is important because the condition of streets and the drainage features near a household may influence household waste handling and disposal practices (Table 4.7).

Table 4.7: Spatial distribution of households by condition of street and drainage feature

Sample distribution by type of street	Sub-Divisional Councils			
	Bamenda I	Bamenda II	Bamenda III	Total
	n %	n %	n %	n %
Footpath	0 0.0	50 20.3	25 12.6	75 16.5
Path usable by wheel barrow / push-push	0 0.0	28 11.5	30 15.1	58 12.7
Loose surface motorable all seasons	4 30.8	117 48.0	63 31.7	184 40.4
Loose surface motorable in dry season	1 7.7	20 8.2	26 13.1	47 10.3
Tarred road in bad state	1 7.7	10 4.1	32 16.1	43 9.4
Tarred road in good state	7 53.8	19 7.8	23 11.6	49 10.7
Total	13	244	199	456 100

χ2-test: χ2=96.99; df=10; P<0.001

Sample distribution by type of drainage feature for households near drainage features

Presence of river	11 100.0	7 9.9	34 73.9	52 40.6
Presence of stream	0 0.0	43 60.6	0 0.0	43 33.6
Presence of standing water	0 0.0	13 18.3	12 26.1	25 19.5
Presence of swamp	0 0.0	8 11.3	0 0.0	8 6.3
Total (valid)	**11**	**71**	**46**	**128 100.0**

χ2-test: χ2=223.97; df=6; P<0.001

No drainage feature: 211 *Source: Household survey, 2013*

Table 4.7 shows the different types of roads and drainage features within the BCC that can influence household waste behaviour. Only a minority of households (20 percent) are located on tarred roads. The majority of households are reached by footpaths, paths used by wheelbarrows and loose surface motorable roads limited for use only in the dry season. Also very prominent are loose surface motorable all season roads to which as much as 40 percent of households access. The paved roads are made up of tarred roads in good condition and tarred roads in poor condition. A statistically significant difference (*χ2-test: χ2=96.99; df=10; P<0.001*) exists in the distribution of the different types of roads in the three administrative subdivisions. Sixty-two percent of households in Bamenda I, compared to 12 percent and 28 percent in Bamenda II and Bamenda III respectively, are accessible by paved roads.

The dominance of unpaved roads has implications on the frequency of waste collection by the council waste collection team, as the trucks and pick-up vans ply mostly the tarred streets of the city and, at best, the motorable all season roads. Consequently, 92 percent, 60 percent and 59 percent of households in Bamenda I, Bamenda II and Bamenda III respectively, can theoretically be accessed by the BCC waste pick-up vans. In reality, the situation is different. The Chief of Service for the Hygiene and Sanitation Department at the City Council explains that waste collection vehicles avoid unpaved roads. This is because most of them are narrow, winding and muddy, especially in the rainy season, making movement difficult. The above results from household responses and information from interview with the council authority give the impression that Bamenda I is in the most favourable position in terms of waste collection.

Table 4.7 also shows that the drainage system includes rivers, streams, swamps and stagnant pools of water. About 38 percent of households are situated close to at least one drainage feature. Spatial distribution by sub-divisional councils shows that there are differences in the number of households that are near one or more drainage features. This difference is statistically significant ($\chi2$-test: $\chi2=223.97$; $df=6$; $P<0.001$). In this connection, Bamenda II and Bamenda III have a higher percentage of households near drainage features (swamps, standing waters) than Bamenda I. This difference can be explained by the physical milieu. Bamenda I is situated on the slope of the Bamenda escarpment, where running water and run-off flows through to the relatively low-lying plains of Bamenda II and Bamenda III. The proximity of such water sources has implications for urban agricultural activities and waste disposal. Water is used especially in the dry season to irrigate farms. In the absence of effective waste collection services within convenient walking range, such bodies of water, particularly rivers and streams are sometimes viewed and used as waste disposal sites by households. It is important to note that, while respondents indicate just one drainage feature, it is very likely that there are households, which are found near more than a single feature.

4.3 Gender Analysis of Household Waste Generation

This section addresses objective two of this study, which focuses on household gender reproductive works and gender division of labour with implications for waste generation. Respondents in the household survey were asked about their perceptions of waste, household activities that generate waste, the type of wastes commonly generated and the roles played by the different members of the household in generating and disposing of waste.

4.3.1 Perception of Waste

Environmental consciousness and waste generation and handling practices including prevention, reuse, recycling, storage and disposal (Littig, 2001) are closely linked to how people perceive waste. Table 4.8 present results for a question that asked how households perceive waste by sex. Findings reveal that 40 percent of men and 37 percent of women see waste as anything useless. A further 23 percent of females and 20 percent of males regard waste as refuse and consider it as whatever is intended for disposal, while 31 percent of males and 37 percent of females are unable to define waste. Such ignorance makes it difficult to appreciate the implications of this category of respondents on waste reduction and handling strategies. This is because some people who cannot give a definition of waste may still handle waste in a responsible way, and some who can give a clear definition may be irresponsible.

Table 4.8: Respondents' perceptions of waste by sex

Perception of Waste	Male		Female	
	n	%	n	%
Anything that is not useful	51	39.8	72	36.9
Whatever pollutes the environment and threatens human health	3	2.3	3	1.5
Residues	7	5.5	11	5.6
Whatever is intended for disposal	30	23.4	38	19.5
Recyclable material	1	.8	3	1 5
I don't know	39	30.5	73	37.4

$\chi 2=0.001$; df=5; P=0.946; $\alpha>0.05$: *Multiple Responses*   *Source: Household survey, 2013*

Analysis by sex shows that there is statistically no significant difference ($\chi 2=0.001$; df=5; P=0.946) between male and female respondents in their perceptions of what waste is. Below are some responses from the code grounding quotation report indicating what respondents perceive waste to be:

"Anything that is not useful" [Female consultant, post-secondary level, aged between 31-35]

"Useless things or anything thrown away and is of no value to the owner" [Male civil servant, postsecondary, "Anything that pollutes the environment" [Female retired worker, postsecondary, aged between 41-45]

"Most dangerous thing in society leading to poor health" [Male, teacher, post-secondary, aged 21-25] aged between 31-35]

"Residual materials from our kitchen, work place etc." [Female civil servant, post-secondary, aged above 50]

"It is any remains of anything which may not be useful to someone but which can be of use to another person e.g. goat dung" [Female Business woman, post-secondary, aged 21-25]

"Remains of articles which not of use at the moment" [Male agricultural technician, post-secondary, aged 41-45]

4.3.2 Analysis of Waste Generating Household Gender Reproductive Works: Activities and Actors

The kind of gender reproductive works (see definition in Section 1.7) performed in the household determines the types and quantities of waste material generated. Consequently, the household questionnaire requested information about major household activities that generate waste, as well as the actors that participate in waste generation. Most households in Africa – unlike the situation in Europe and North America, where single person households and single parenting are common—are extended families in which all four types of persons shown in Table 4.9 are present. Therefore, specific data were not collected about whether all the roles were represented in each household, and no attempt has been made to construct separate tables for situations where one or more of the roles were not represented. Indeed, most such tables would have included too many persons for analysis. The results in Table 4.9 must therefore be interpreted with the caution that things might be rather different for households where not all roles are represented For example, where there are no children, then adults must do all of these things even though children would do some of them if there were.

Table 4.9: Household waste generation: distribution of responses by gender of actor

Household waste generation activities	Involvement of actor in various waste generating activities							
	Adult Female		Girl Child		Adult Male		Boy Child	
	n	%	n	%	n	%	n	%
Food preparation	273	80.5	121	35.7	42	12.4	37	10.9
Sweeping	96	28.3	204	60.2	72	21.2	134	39.5
Compound clearing/yard trimming	65	19.2	76	22.4	179	52.8	142	41.9

N=339; χ²=124.15; df=6; P<0.001. N.B. *Multiple Responses. Source: Household survey, 2013*

Table 4.9 shows the involvement of adult females, adult males, girl children, and boy children in three major waste generating activities; food preparation, sweeping of the domestic environment, and compound clearing. The results suggest that a female adult and a girl child participate in food preparation in 81 percent and 36 percent of households respectively. Relatively few households have an adult male (12 percent) or a boy child (11 percent) participating in this activity. Where sweeping is concerned, an adult female and a girl child take part in 28 percent and 60 percent of households respectively, compared to 21 percent and 40 percent for an adult male and boy child. Evidently, children, especially girls, are often assigned this task. Compound clearing, on the other hand, is dominantly a male activity. In 53 percent and 42 percent of households surveyed, an adult male and a boy child respectively participate. This is considerably higher than the 19 percent and 22 percent participation level for adult females and the girl children. Overall then, household activities that generate waste are not completely sex specific; persons from all four categories participate, at least occasionally, in all activities. Rather, these activities are gender related, i.e., they are more frequently done by one gender than another, and defined by role expectations that associate certain functions with either being male or female.

4.3.3 Frequency of Household Solid Waste Generation Activities

The frequency with which activities with waste generation potential are carried out can also influence the quantity and nature of waste that households generate. Consequently, the survey questionnaire included an item about how often household activities with potential to generate waste are performed: daily, bi-weekly, weekly, bi-monthly and monthly (Table, 4.10).

Table 4.10: Distribution of responses by frequency of waste generation activities

Waste generation activities	Daily		Bi weekly		Weekly		Bi- monthly		Monthly		Undefine	
	n	%	n	%	n	%	n	%	n	%	n	%
Food preparation	284	83.8	32	9.4	9	2.7	1	0.3	0	0.0	13	3.8
Sweeping	254	74.9	55	16.2	20	5.9	1	0.3	1	0.3	8	0.0
Compound Clearing	10	2.7	32	9.4	91	26.8	62	19.8	116	34.2	23	6.8

N=339   Multiple Responses. *Source: Household survey, 2013*

Table 4.10 shows that an estimated 84 percent of households carry out food preparation activity daily, nine percent prepare food about twice a week, and three percent about once a week. Sweeping of the domestic environment is done daily in about 75 percent of households; in 16 percent sweeping is performed bi-weekly and in six percent about weekly. For compound clearing, results indicate that it is performed daily in three percent of households, in 20 percent it is done bi-monthly, in 27 percent it is carried out weekly, and in 34 percent it is done once a month. In seven percent of households compound clearing is never done. This last category of households could be a reflection of those that live in buildings divided into apartments or those without spaces for any form of vegetation or gardening. It is important to highlight here that compound clearing, the only male dominated task, is a much less frequent activity. This is the opposite of food preparation, which is the most frequent and the most female typed activity. The disparity in the male and female level of involvement in household gender reproductive works reflects gender division of labour, which is influenced by societal norms. Therefore, one can safely conclude that female gender related activities tend to generate much more household waste than male related ones.

4.3.4 Gender analysis of the level of participation in waste generating household reproductive work

To determine which household actors contribute the most to waste generation, a question was included that asked for more detail about how often people in various roles participate in each activity. Table 4.11 shows the extent to which men, women and children are involved in household waste generation activities by examining how frequently they participate in these activities. The categories "always," "sometimes" and "never" are used to determine the level of involvement.

In 89 percent and 16 percent of households, the adult female and adult male respectively always take part in food preparation activities. On the other hand, the adult female never participates in only about two percent of households compared to 41 percent for the adult male. Making reference to Table 4.10 (Section 4.3.3), which indicates that 84 percent of households carry out food preparation daily, and it is the most frequent household activity that generates waste, we can dare to compare the extent to which one gender generates more waste over the other. In 89 percent of households (Table 4.11), the adult female always takes part in food preparation when compared to 16 percent of households where the adult males always participate. We can therefore estimate that an adult female possesses 5.6 percent times more potential to always generate waste from food preparation than the adult male. The result also implies that the adult male has a 20.5 percent (Table 4.11) higher chance than the adult female never to generate waste from food preparatory activities. Focusing on children's participation, data show that, in 61 percent and 13 percent of households, the girl child and boy child respectively, are always involved in food preparation activities, while in two percent and 32 percent of households the girl and the boy child never participate. These results point to the fact that, the girl child has the potential to always generate food waste 4.7 times more than the boy child does. More so, the boy child has 16 times the chances of never generating food waste, as the girl child will do (Table 4.11). It can therefore be concluded that adult females and girl children are much more involved in generating food preparation waste than adult males and boys.

Table 4.11: Household waste generation actors: Distribution by level of involvement

Household Activities		Level of involvement by gender of actor in waste generation activities											
		Always				Sometimes				Never			
		AF	GC	AM	BC	AF	GC	AM	BC	AF	GC	AM	BC
Food-	n	270	87	17	11	28	52	45	46	5	3	43	27
Preparation	%	89	61	16	13	9	376	43	55	2	2	41	32
Sweeping	n	73	152	37	94	63	38	48	53	4	6	29	7
	%	52	78	33	61	45	19	42	34	3	3	25	5
Compound-	n	44	49	87	81	48	42	64	58	16	16	15	7
clearing	%	41	46	52	56	44	39	39	40	15	15	28	5

N=339. χ^2=113.98; df=6; P<0.001   Source: Household survey, 2013: AF-Adult Female; GC-Girl Child; AM-Adult Male; BC- Boy Child

Considering the activity of sweeping, responses indicate that, in 52 percent of households, the adult female, as compared to 33 percent of the adult males; always take part in this activity. On the other hand, three percent and 25 percent of adult females and males respectively are never involved. Consequently, the adult female has about 1.6 percent more chances of always generating waste from sweeping than the adult male does. On the contrary, the adult male possesses about 8.3 percent chances of never producing waste from sweeping than the adult female. Focusing on children's participation, results indicate that in about 78 percent and 61 percent of households, the girl child and boy child respectively are always involved in sweeping. Sweeping of the domestic environment shows an active participation of children with a slight lead by the girl child. There is generally a relative upward trend in male participation (adult male and boy child) in the activity of sweeping than in food preparation.

Focusing on participation in compound clearing by gender of the actor, findings show that in 41 percent and 52 percent of households the adult female and adult male are always involved. In 46 percent and 56 percent of households, the girl child and the boy child always participate. Adult male and boy child dominance and an active representation of the adult female and girl child are observed. It is however important to note that though compound clearing is considered a male dominated activity, the figure of 53 percent adult male participation falls far below the 89 percent adult female participation in food preparation which could be rated as women's related type activity. It is only in 16 percent of households that adult males are always involved in food preparation, which is considered women's work. This figure for male participation is relatively small when it is observed that in 41 percent of households, the adult females are involved in compound clearing which is considered men's work.

Overall, the results show that a significant association ($\chi^2=124.15$; df=6; P<0.001) exists between the level of involvement in household waste generating activity and the gender of the actor involved. The obvious gender division of labour manifested in the results is probably the result of cultural norms (Moser, 1993) common to most African societies. The results presented and analysed make very visible the central role of women in household waste generating activities and the conclusion that household waste generation is both gendered and feminised.

4.4 Types and Quantities of Household Solid Waste in BCC

Household waste includes organic (biodegradable) and inorganic (non-biodegradable) materials. The term organic waste is understood in this research as waste with high biodegradable content that tends to decay quickly. Most household organic waste comes from food products and compound/yard clearing. Inorganic waste includes material with low or slow rates of decay. For this reason paper and cartons, although technically biodegradable, are classified here as inorganic (non-biodegradable) waste. Results from household on-site waste measurements in the dry season for one week (Table 4.12) are used to determine average waste quantities and classify waste as organic and inorganic. The results showing the average quantities presented in Table 4.12 are from four neighbourhoods of varied income levels chosen to ensure a diverse social class and spatial representation. The different income class quarters are: Foncha Street (high income), Ndamukong (middle income), Old Town (low income) and Abangoh (peripheral). The peripheral zone has no distinct income level, but it is important as a peri-urban zone, where the traditional lifestyles of the indigenous population have implications for waste amounts and waste types. Table 4.12 shows descriptive and comparative statistics for amount and types of waste generated by 109 households chosen for on-site measurements.

The mean, median and standard deviation in the weight of waste generated by all three neighbourhoods combined are 25.9kg, 18.0Kg and 23.3Kg respectively. The maximum average weight generated a week by a household is 117.0Kg and the minimum is 3.0Kg. There is a statistically significant difference ($\chi2=14.10$: P=0.03) in the weight of waste generated across the four neighbourhoods. The total amount of waste in weight generated in the high income and middle-income class neighbourhoods of Foncha Street, Ndamukong, and Old Town are 27.9Kg and 20.1Kg, and 18.6Kg respectively, suggesting that total waste volume decreases with neighbourhood income. Location appears to be a more important factor, however.

108

The peripheral neighbourhood of Abangoh generates 38.7Kg of waste weekly, far more than any of the others. The difference is significant at the .01 level. It is important to note that disposal method and perception can also affect waste quantities dumped.

Table 12: Descriptive and comparative statistics for amount (weight) and types of waste generated by neighbourhoods

Waste quantities in weight		Neighbourhood					Kruskal Wallis Test
		Foncha Street	Ndamu-kong	Abangoh	Old town	Total	
Kilograms of organic waste per household per week	N	38	28	20	23	109	χ2=17.62
	Mean	18.18	12.68	29.05	13.52	17.78	
	Median	11.00	8.00	27.50	12.00	12.00	
	Std. Deviation	19.01	11.67	17.00	7.07	15.91	P=0.01
Kilograms of inorganic waste per household per week	N	38	28	20	23	109	χ2=2.17
	Mean	9.71	7.39	9.60	5.09	8.12	
	Median	4.00	3.00	6.00	4.00	4.00	P=0.54
	Std. Deviation	16.40	14.68	10.85	3.90	13.16	
Total Kilograms of waste generated per household per week	N	38	28	20	23	109	χ2=14.10
	Mean	27.89	20.07	38.65	18.61	25.90	
	Median	17.50	11.00	30.50	17.00	18.00	P=0.03
	Std. Deviation	27.79	20.50	25.05	7.69	23.32	

The mean median and standard deviation for organic waste in all four neighbourhoods combined are 17.8Kg, 12.0Kg and 15,9Kg. Once again, waste generation increases with neighbourhood income, but Abangoh, the peripheral neighbourhood, generates the most,

29.0Kg, which is more than double the weights for Ndamukong (12.7KG) and Old Town (13.5Kg).The difference is again statistically significant. The high level of organic waste generation for Abangoh can be attributed to the characteristics of the population, which is mostly indigenous and tends to have a strong attachment to traditional meals. Traditional meals have an especially high potential to generate waste from their peelings and packaging. These include dishes such as "achu" (a pounded mixture of boiled cocoyam and green banana), corn products, cassava, yams, plantain and green banana and accompaniments (vegetables). Also included are wastes from eggs, potatoes, sugar cane, corn, and beans, as well as peelings from numerous kinds of fruit. In addition, this is a quarter with much open space with grassy vegetation that needs clearing. Composting of organic waste at the domestic and community levels in this area could therefore be a promising option that could greatly reduce the amount of waste that has to be transported. It would also reduce the amount of waste that goes to the landfills, thereby extending the life span of landfills and reducing collection and transportation cost for the Council.

The mean, median and standard deviation of inorganic waste (recyclable and non-recyclable) generated per household in all four neighbourhoods combined are 8.1Kg, 4.0Kg and 13.2Kg. Although there is a slight tendency for rich neighbourhoods to produce more inorganic waste, the difference in the average weight of inorganic waste generated in the different neighbourhoods does not approach statistical significance, indicating that the differences in overall waste weight are due primarily to differences in organic waste generation Table 4.13.

The average (mean) weight of waste generated daily per capita in all four neighbourhoods combined is estimated at about 0.71Kg. There is a statistically significant difference ($\chi2=17.27$, P=0.01) in average weight of waste generated per capita per day in the different neighbourhoods/quarters. Residents of Foncha Street generate about 0.9Kg; Ndamukong, 0.4Kg; Abangoh, 1.0Kg and Old Town, 0.6Kg. As observed in the previous table, the largest difference is between the peripheral neighbourhood and the others. For organic quantities (biodegradable) waste generation, the average weight generated daily is .51Kg per capita. Abangoh again generates the highest amount daily in terms of weight (.71Kg per capita). Once again, there was no significant difference among the neighbourhoods in inorganic waste generation.

Table 4.13: Household daily waste generation quantities and qualities

Description of waste quantities and types		Waste quantities by quarter					Kruskal Wallis Test
		Foncha Street	Ndamu-kong	Abangoh	Old Town	Total	
Kilograms of organic	N	38	27	20	23	108	$\chi 2=19.64$
waste per capita daily	Mean	.58	.27	.72	.44	.50	P=0.01
Kilograms of inorganic	N	38	27	20	23	108	$\chi 2=4.11$
waste per capita daily	Mean	.30	.12	.23	.17	.22	P=0.25
Total Kilograms of waste	N	38	27	20	23	108	$\chi 2=17.27$
generated per capita daily	Mean	.88	.38	.95	.61	.71	P=0.01

Source: Household survey, 2013

In addition to food preparation, sweeping of the domestic environment and compound clearing, other activities especially income generating ones, carried out in some households contribute to waste generation. Consequently, the questionnaire included questions about business related activities carried out in households and their immediate environs. The results indicate that some form of small-scale business activity is carried out in a majority (92 percent) of the households. The most common include the production of pastries, milk products (e.g. yoghurt and ice cream), local beverages (e.g. corn beer), fruit juice, artisanal products (e.g. pottery, weaving, stitching), general repairs, and poultry for sale.

The business related activities generate waste from both the production process itself and the materials used for packaging. The specific waste types reflect the type of raw material used and packaging. Of the 311 households that carry out some form of production for commercial purposes, 36 percent (112), use some form of packaging. Out this number that use packaging, nine percent use textile and fibre; 14 percent use paper and/or carton; 71 percent use plastics, six percent use glass, and other material. In terms of the types of solid waste that their commercial activities generate, 25 percent reported biodegradable waste, for example from foodstuff, sawdust, and dung from poultry and piggery. The 75% who reported that their activities (both production and packaging do not produce biodegradable waste generate non-biodegradable waste in the following proportions: 34 percent plastics, 21 percent paper and cartons and others 20 percent (stones, tiles, glass, textile and metals).

Field observation and researcher's life experiences reveal that there are variations in the daily and seasonal waste production amounts. On-site household waste measurements carried out in 108 households from four different socioeconomic class neighbourhoods/quarters (Foncha Street, Ndamukong, Abangoh and Ntambag (Old Town) show that greater waste quantities in weight and volume are produced during weekends, especially on Saturdays and Sundays, when more cooking and other domestic cleaning activities are carried out. Existing literature also suggests that waste types and quantities vary with changes in season.

Though this study did not carry out measurements in both the dry and rainy seasons, other studies have been carried out in other tropical areas with climatic conditions similar to Bamenda. For example, Ngnikam (2000) and Tanawa et al. (2002), in waste studies in Yaounde and other cities in the central African region respectively, suggest that per capita weight generated might be as much as 29-39 percent higher during the rainy season. The average difference in the moisture content of waste between the wet and dry season is about 58 percent, with relatively higher amounts registered in the wet season. The rainy season (April to October) is usually characterised by harvesting. This is the time when fruits ripen and high waste generating products, such as leafy vegetables and other crops like maize and beans, are available in abundance. It is also the time when waste from yard cleaning increases because of frequent trimming of hedges and grass. The general increase in waste quantities and types is thus more applicable to biodegradable waste than non-biodegradable.

In summary, findings in this chapter have addressed objectives one and two of this study, which sought to describe the demographic profile of households and the waste generation potential of various household waste activities and actors by gender. Major findings concerning objective one include the following: Almost all household respondents and decision makers of both sexes in the BCC have at least a minimum level of formal education. Decision makers in the household who are here referred to as fathers and mothers are involved in different activities of the different sectors of the economy. In terms of income levels, as much as 87 percent of households earn less than 200,000FCFA (about $400USD) a month, indicating a relatively low purchasing power of the City. In terms of the natural environment where households are located, over one-third of them are found near a drainage feature – river, stream, swamp, and gutter –, which can influence household waste disposal practices. Over 75 per cent of the roads within the municipality are unpaved, and this situation can have an impact on waste collection and transfer, as it is difficult for trucks to operate on unpaved roads, which are sometimes in very bad state. Gender analyses of respondents' perception of waste reveal that there is statistically no significant difference in the way men and women perceive waste. More than half the number of respondents (66 percent) view waste as anything that has no value. Over a third says that they do not know how to define waste. Waste as a possible resource was hardly articulated by respondents.

In terms of household activities that have potential to generate waste, findings identify food preparation, sweeping of the house, and compound clearing as the most important. Resultantly, the organic waste amounts in weight surpass that of the inorganic. Women, men and children all participate in these activities that can generate waste, but with different levels of involvement in different activities. Both adult females and girl children take very active part in household tasks that are considered women's roles. In addition, they are also sometimes very involved even in male perceived household tasks as compound clearing. It can therefore be concluded that both sexes participate in waste generation, but women account for more of it. The level of participation is evidently guided by the household gender division of labour, which in itself is determined by societal norms and traditions.

Chapter Five

Gender Analysis of Household Solid Waste Management

Research findings presented in this chapter are related to the key activities and actors in household waste management practices both overall and by gender. Results come from the household survey questionnaire, which investigated the level of involvement of men, women and children in waste storage, separation, reuse, transportation, and disposal. The goal is to identify the actors with influence in household waste management activities. These are the actors whose degree of participation in household waste handling activities affects the effectiveness of the Bamenda City Council (BCC) waste management policies and strategies. Conversely, household waste handling activities, workload and even livelihood sources can be affected by the BCC waste management strategies.

5.1 Household Solid Waste Storage Situation

The conditions under which waste is stored have the potential to influence the hygiene and sanitation of the domestic environment and the extent of waste reuse and separation. In this section, results from survey questions about the types of waste storage containers used and the average length of time waste is stored are reported. According to the survey results, a variety of containers made of different materials are used to collect and store waste in households. In 94 percent of households, waste storage containers made of non-biodegradable material are used. In seventy percent of households in which non-biodegradable waste containers are used, the containers are generally made out of plastic materials in the form of bags, buckets and baskets.

The household survey questionnaire also inquired about the average size of the waste containers used by households, as the frequency with which the Council collects waste may affect the size of the containers used. The study also sought to determine whether there is a relation between size of container and household size and the frequency of waste disposal. Results reveal that the average size of household waste storage containers is equivalent to 1.7 sac and motto bag (about 34 litres). Using the Spearman's rho, there is no correlation (Spearman's rho: r=0.060; P=0.372) between the size of waste container and frequency of councils' collection in the different quarters/neighbourhoods as discussed in chapter six in detail. Not surprisingly, there is a negative correlation (r=-.159; P=.008) between waste storage container size and how often households empty their waste. There is a positive relationship between size of waste storage container and amount in weight and volume of food waste generated. This relationship is statistically significant (r=.460; P=.000), implying that the greater the amount of food waste generated, the bigger the storage container. The relation is same for non-biodegradable waste materials, such as plastics, paper, cartons and glass r=.251; P=.028). There is also a positive relationship between household size and size of waste storage container r=.114; P=.062). Though relationship is not quite statistically significant, it suggests that the bigger the household size, the larger the waste storage container and vice versa. This trend of results is expected.

The duration within which households store their waste can influence the presence of disease vectors such as flies, cockroaches, rodents and even mosquitoes. Consequently, a further question was posed to household respondents as to whether or not their waste storage containers had lids. Results show that 30 percent of households have waste storage containers with lids, while 70 percent do not. Curiously, the possession of lids does not guarantee their regular usage. Fifty-six percent of the households that report having lids for their waste containers indicate that they cover their containers regularly, while the rest do not. Absence or negligence in the use of waste storage container lids can attract domestic animals like dogs, cats, and fowls, as well as rodents that rummage through containers. Bearing this in mind, respondents were asked if their households have experienced the following problems around their waste container: presence of dark flowing water, odour, mosquitoes/flies, cockroaches and domestic animals. Table 5.1 presents the sanitation situation around waste storage containers in households with risk potentials.

Table 5.1 Sanitation condition of waste storage container

Characteristics	Always			Sometimes		
	Bda I	Bda II	Bda III	Bda I	Bda II	Bda III
Dark Flowing Water	33%	26%	24%	67%	74%	76%
	(3)	(55)	(24)	(6)	(157)	(77)
Odour	27%	38%	40%	73%	62%	60%
	(3)	(81)	(42)	(8)	(130)	(63)
Mosquitoes /flies	27%	38%	24%	73%	62%	76%
	(4)	(71)	(44)	(7)	(141)	(62)
Cockroaches	40%	28%	20%	60%	72%	80%
	(4)	(59)	(20)	(6)	(152)	(81)
Domestic Animals	55%	27%	32%	46%	73%	68%
	(6)	(55)	(33)	(5)	(150)	(69)
Rats	30%	38%	29%	70%	62%	71%
	(3)	(75)	(30)	(7&	(123)	(73)
Scavengers	50%	49%	46%	50%	51%	54%
	(5)	(93)	(44)	(5)	(97)	(52)

Multiple Responses   **Bda=Bamenda**   *Source: Household Survey*

Table 5.1 shows that almost all households always or sometimes experience some form of inconvenience resulting from the way waste is stored. Households are more likely to identify with all of the problems occurring "sometimes" rather than "always." No household indicates the total absence or rare occurrence of any of these nuisances. Results indicate that some problems may be more pronounced in some areas, but on the overall, they are widespread across the city and one area does not seem worse than the others do. For example, in Bamenda II, about 26 percent and 74 percent of households always or sometimes have dark flowing water from their waste storage containers. The risk of having odour is always or sometimes present in 38 percent and 62 percent of households. In 38 percent and 62 percent of households respectively, cockroaches are always or sometimes seen. Domestic animals, rats and scavengers are also always or sometimes visible in all households. The results in Bamenda III and I reveal similar trends as those of Bamenda II.

5.2 Household Waste Separation

Waste separation is known to facilitate waste minimisation efforts by making reuse and recycling, scavenging, waste treatment and consequently disposal less difficult; however, waste separation at the level of the household is generally not typical of urban dwellers in African cities (Sotamenou, 2010; Medina, 2008, Fombe, 2007, Achankeng, 2005). In this section, research findings from a survey question that solicited information about household waste separation practices and the actors involved are presented. The results revealed that more than half (59 percent) of households within the city do not separate their waste, and there is no statistically significant difference among the different subdivisions of Bamenda I, II and III ($\chi2=2.168$; df=2; P=0.338). The difference in the percentage of respondents willing to separate waste from all three subdivisions in the City is not statistically significant ($\chi2$-test: $\chi2=2.495$; df=2; P=0.287. Interestingly, there is no significant relationship between level of education of household decision makers and waste separation. The fact is that 59 percent of households that do not separate waste present a very important obstacle for any waste management plan that envisages treatment, recycling and/or composting and even scavenging at macro level. In households in which sorting is done before storage or disposal, waste is generally separated into biodegradable and non-biodegradable components. In some households, only waste with monetary value is removed.

5.2.1 Household Waste Separation by Gender

Household gender division of labour influences the contribution of different categories of household members to separate waste. Table 5.2 presents results concerning the level of participation of household members in waste sorting activities by gender (adult female, adult male, girl child and boy child).

Table 5.2 Level of participation in waste sorting by gender of actor in households that separate waste

Gender of Waste Sorter	Level of involvement							
	Always		Sometimes		Never		Total	
	N	%	n	%	n	%	n	%
Female Adult	88	69.3	31	24.4	8	6.3	127	100
Girl Child	27	26	57	54.8	20	19.2	104	100
Adult Male	37	34.3	33	30.6	38	35.2	108	100
Boy Child	22	21.6	45	44.1	35	34.3	102	100

Source: Household survey, 2013

Data show that in 69 percent of households, the adult female is always involved in waste separation activity as against 34 percent where the adult male always participates. Everything being equal, this result indicates that the adult female has the tendency to participate twice as much as the adult male. The difference in the proportion of households, 26 percent and 22 percent, in which the girl child and boy child always participate in waste is not very remarkable.

Results further indicate that it is only in six percent of households that the adult female does not participate in waste separation. Contrary to this small proportion of households were women do not participate in waste separation, as much as 35 percent of households indicate that the adult male does not take part in this task. Examining the level of non-participation of children, 19 percent and 34 percent for the girl child and boy child respectively are reported. These proportions clearly suggest that adult females and girl children are more involved in waste sorting activities than do the adult males and boy children. This remarkable level of involvement points to the fact that any waste management strategy intended to be effective must involve the producer and the sorter especially at the primary stage. In this case, the female adult is the main generator and evidently the main sorter of household waste. This result affirms that household waste management is feminised. Consequently, any municipal initiative aimed at handling household waste must thus reflect this gender specialization in household waste generation and sorting.

5.2.2 Reasons for Household Waste Separation

To further investigate waste separation, the survey questionnaire included a question about the reasons why the households that practice waste separation do so. The responses most frequently given allude to soil enrichment, degradability, nature of waste, injury potential, and usage/recycling, feed for animals and flammability (Appendix VII). Forty-two percent of households separate waste in order to use the organic material as manure. Another 44 percent do so with the intention to make sure that non-degradable waste is disposed of separately, and to avoid smell that may be generated by some waste. Four percent of households separate waste to gain feed for pigs and dogs and ten percent to burn flammable waste such as paper. Below are some responses from the code grounding quotation report suggesting reasons for waste separation:

"It goes mostly to the farm" [Male, retired worker, aged above 50]

"To burn plastics and throw which can decay in the farm" [Female teacher, post-secondary, age above 50]

"In order to burn plastics" [Male driver, primary level, aged above 50]

"To separate biodegradable from non-biodegradable" [Female teacher]

120

"Some easily get rotten" [Female, secondary, aged 31-35]

"Because some waste may be in dust form" [Female, businessperson, post-secondary, aged 16-20]

"Some is water and may soak others" [Female business woman, post-secondary, aged 21-25]

"Some of the waste undergoes decay fast" [Male, post-secondary, aged 36-40]

"To avoid waste that can harm someone" [Female student, postsecondary, aged 26-30]

"To keep away flies and mosquitoes" [Adult Male businessperson, post-secondary]

"Because of broken glasses" [Adult male journalist, primary education]

"Because if some are mixed it will be poisonous" [Male driver, secondary, aged 31-35]

"Avoid odour" [Male farmer, primary, aged 26-30]

"Some at times are still useful" [Female traditional Doctor, primary education, aged 41-45]

"Because I use the plantain peals for niki (local potash)" [Female, farmer and businessperson, primary, aged above 50]

"Some are used as food for pigs" [Female, post-secondary, trader, aged 21-25]

"To burn paper" [Housewife, secondary, aged 36-40]

"So as to burn paper that does not decay" [Female businessperson, primary, aged above 50]

"Some wastes are burned while others are thrown in the farm" [Adult female trader, post-secondary]

5.2.3 Reasons for Non-Separation of Household Waste

Fifty nine percent of respondents indicate their households do not separate waste. Diverse reasons are advanced for this inaction. A large proportion (42 percent) of these respondents does not see sorting waste important. About a quarter (23 percent) of the respondents, consider it time consuming. Another 23 percent indicate that they lack access to farmland where decomposable material could be used as manure. About a tenth (11 percent) perceives sorting as burdensome and inconvenient, and seven percent view it is as council responsibility. Here are some sample responses from the code quotation grounding report:

"I see no need" [Female, post-secondary, aged between 31-35]

"No need sorting since we dump everything" [Female, taxation officer, post-secondary, age between 36-40]

"Because I consider waste as waste; there is no need for separation" [Male, office worker, post-secondary, aged 31-35]

"It is a mere waste of time" [Male civil servant, aged 36-40]

"It is waste of time" [Male, no occupation, primary level, aged 16-20]

"There is no need because the collectors do not consider sorting" [Female teacher, post-secondary, aged 46-50]

"Waste is waste, so we cannot separate them" [Male builder, primary, aged 16-20]

"Because anything rejected is waste" [Female decorator, Primary, aged above 50]

"Because there is no machine to sort waste" [Male, civil servant, post-secondary education, aged 31-35]

"It has odour, hence difficult to sort" [Female homemaker, secondary, aged 31-35]

"Because I don't have another storage container" [Male nurse, post-secondary, aged 36-40]

"Because all of them are waste" [Female, aged 26-30]

5.2.4 Motivating Factors for Household Waste Separation

Findings in this section address a question posed to respondents about what could encourage or oblige households to separate waste. Respondents of households that do not separate waste suggest varied conditions and incentives that might spur them to do so. Fifty-six percent of respondents link their responses to the need to be educated on waste issues and effective presence of council waste service. Another 19 percent declare that they will separate waste on condition that there are persons ready to receive waste material for further use. An additional three percent suggest the presence of recycling industries is required. Other responses mentioned the provision of separate waste containers for different types of waste by the council; the need to attribute economic value to household waste material; and the presence of farmland on which organic waste can be used as manure. Some suggest that they will practice waste separation if others do it. The difference in the percentage of respondents willing to separate waste from all three subdivisions in the City is not statistically significant (χ2-test: $\chi2=2.495$; df=2; P=0.287. Below are some typical responses suggesting incentives that might motivate household respondents to separate their waste:

"If the council encourages sorting" [Female teacher, post-secondary, 46-50]

"A clear explanation" [Male painter, secondary education, aged 26-30]

"Sensitisation on health hazard of waste" [Male doctor, post-secondary, 36-40]

"Storage containers" [Male driver, post-secondary, aged 21-25]

"If council collects them separately" [Male consultant, postsecondary, aged 41-45]

"If they carry it every day" [Adult male journalist, primary]

"If I have a helping hand" [Female public service worker, post-secondary, aged 36-40]

"If it has use to one" [Female teacher, aged above 50]

"Placing bins around" [Male driver, primary level, aged above 50]

"Avoid odour" [Female teacher, post-secondary, aged 26-30]

"Recycling industries if they are set up" [Male civil servant, post-secondary education, aged 41-45]

"If I have a farm to deposit compost waste" [Female teacher, post-secondary, aged 36-40]

"If other people do the same" [Female call box, post-secondary, aged 41-45]

"Nothing because it is necessary' [Male electrician, postsecondary, aged 21-25]

"Only payment can motivate me" [Male teacher, post-secondary, aged 25-26]

"Easy to burn" [Female police officer, aged 21-25]

5.3 Waste Collection and Transportation

Waste collection is generally considered at two levels; primary waste collection and secondary waste collection. Primary collection or pre-collection is concerned with assembling, storing, sorting and transferring of waste from generation site to treatment facility or secondary collection site by household members or paid efforts. Secondary collection occurs at the point where waste is amassed (for example the communal bin/site) and transported for final disposal by the council. In this section we address primary collection focusing on how often household waste is removed and transferred to the waste collection site and who does it. Survey results show that 35 percent of households carry out waste daily, 19 percent do it once every two days, 30 percent do so every three days, and the rest once a week or less. Multiple factors are likely to influence the frequency of waste removal from home, including distance to the disposal site. Small waste storage containers also encourage the frequent disposal of waste. Waste materials that could be considered as nuisance by emitting pungent smell, leachate, attracting animals and flies and other insects are also probably disposed of more frequently.

To determine the degree of involvement of different members of the household in waste removal, a gender analysis is done. Caution is drawn to the fact that no adjustment on whether all categories of household members (adult female, adult male, girl child and boy child) are represented in every household. It is assumed as earlier mentioned, that it is likely that most households in an African setting will contain them all, and therefore results can be generalised. The results show that all types of household members at one time or another transfer waste from home to primary or secondary disposal site. However, the level of participation of each gender in primary waste transportation varies with different households. Survey data respond to a question that solicited information on who moves waste from the household to the public waste collection site. The descriptions of always, sometimes and never were given as options. Research results show that in 60 percent of households, children are those always involved in transferring waste away from domestic environments. Further detail reveals that the boy child of the children category takes the lead role in waste transfer from the house, and more so boys who are below the age of 16. This category is followed by the adult female/mothers (30 percent) and eight percent by paid transporters. It is only in two percent of households in which adult men (generally considered heads of household) are involved.

5.4 Waste Disposal, Sensitization and Environmental Consciousness

This section presents research findings about the level of environmental awareness of households. The level of environmental consciousness is assessed from the household waste disposal practices, and the level of education of respondents on waste management issues. Attention is given to the differences in the levels of environmental consciousness and environmentally friendly behaviour of men versus women. Such information is useful for waste operators and environmentalist in identifying their target population in the bid to create environmental awareness strategies.

5.4.1 Household Primary Waste Disposal Sites
Results in Table 5.3 and Plate 5.1 address the question that required information on the most used disposal site by households.

Plate 5.1 Illegal Waste Disposal Site below Foncha (Nkwen)

Data reveals that households use both council provided waste disposal sites and sites they select on their own. Council provided waste sites/facilities include mobile waste collection pick-up van and the stationary public trash cans/public skip. The privately or individually determined sites include roadsides and gutters, open spaces inside and outside of compounds, composting/farms, and rivers/valleys/streams. In some instances, waste is disposed of in runoff produced during heavy rains. These poor disposal methods end up blocking drains and stream channels, leading to floods, which occur often in the Nki Bahseu and Lipacan in the Musang and Nkwen areas.

Table 5.3 shows that only 32 percent of households use council provided facility, while a striking 38 percent dispose of their waste in valleys, river, streams, open spaces, roadsides and gutters. Another 30 percent dispose waste within the confines of their compound by throwing in the yard to serve as manure, or by burning. Waste material from certain products like cocoyam and plantain peelings and leftover of food that can serve as animal food is sometimes offered to persons who need and are ready to collect them.

Table 5.3 Household primary waste disposal sites

Where waste is disposed	Count (n)	Percentage (%)
Mobile (itinerant) waste van	67	21.3
Public skip (dust bin)	33	10.5
Valley/river/stream	22	7.0
Road/street sides/gutters	73	23.2
Open spaces out of compound	25	8.0
Open space in compound (yard)	18	5.7
Pit in compound	35	11.1
Compost/farm/feed/offer	41	13.1
Total	**314**	**100**

Source: Household survey, 2013

Experience and observation indicates that littering is also another widely used waste disposal practice. Littering is the careless throwing of waste on the ground around houses, offices, market places, streets, public places and parks. Such wastes include nutshells, wrappings from sweets and biscuits, cigarette butts, fruit peelings, plastic bottles, plastic bags, etc.

The justifications for disposing waste at selected sites vary with the facility used. The dominant reasons given are; the most available facility, accessibility and convenience. Some peculiarities however exist. Forty-two percent who use the gutter say runoff from rainfall washes waste away, and some add that there is no threat of sanction or penalty. Those who dump waste on roadsides say they are influenced by the action of others. In addition to attributes of proximity, availability, accessibility and convenience, those who use the public bins sometimes, add that they have no farm or yard. Some say they use the public bin because they do not want to burn papers. Others add that, they use the public bin to avoid odours around the home environment. Still, some declare that they use the public bin because it is the only authorized site. Households that use the yard and farm or make compost say the compost is meant to serve as fertilizer.

In their responses to a question about what households do with the biodegradable waste from food, leaves and yard trimmings, a substantial majority (77 percent) indicate that they dispose of it in the same way as all other waste. Another 13 percent suggest that such waste is used to make compost and for other farm related needs. This proportion that does composting is surprising for an urban population, which would not be expected to be involved in agricultural activities. However, one would have expected much more for households in Bamenda where the peripheral zones are most cultivated by the population that lives in the city. However, the people of the peripheral zone of Abangoh do so. This is expected, as there is available open space. According to the respondents, many factors influence household waste composting. An overwhelming 94 percent of households that carry out composting do so in order to produce fertilizer. A few respondents say it is intended to reduce the amount of waste to be transported to the public waste collection site. Others indicate that their households do composting because there is space available and council fails to collect waste.

Varied reasons are advanced by households where composting is not done. About 25 percent of households indicate that they lack space, 22 percent do not perceive the need for composting and 18 percent are constrained by time. Another twelve percent of households do not have access to farms where compost could be used. The rest indicate that they have ready access to an authorized public waste skip or dumping site. Some also argue that composting is inconvenient and nasty, and others say that the waste quantities they generate are too small to justify composting.

5.4.2 Waste Sensitization and Environmental Consciousness

Environmental awareness and knowledge about waste management is obtained from diverse sources and is the responsibility of various authorities. Waste managers and related authorities have a responsibility to educate the public about waste handling issues in order to promote efficient management and reduce harm to the environment. This section presents findings from a question about households' knowledge of waste management and its effects on the environment and the sources of their information. The results reveal that 79 percent of respondents have received no education at all about waste and waste management issues. The respondents who have had some form of education identify varied sources: media (37 percent), school milieu (19 percent), community education and public information campaigns (16 percent), health centres (six percent), church (eight percent) and others including home environment, researchers, seminar, books and discussions (14 percent).It is however important to observe here that there is a high chance that a respondent could be educated by more than one source.

A further question inquired about the agents of environmental education from whom respondents could have gotten awareness on waste and environment related issues. Most frequently mentioned are teachers (27 percent), journalists (12 percent), health team members (eight percent), Non-Governmental Organisations and Community Initiative Groups (seven percent), and the waste collection team from the council (eight percent). Other actors mentioned include social group members, friends, family members, the Regional Delegation of the Ministry of Environment, mothers, church members and market managers. It is striking and disappointing that the Council accounts for only eight percent of waste management education, as they are the lead managers of waste issues within the municipality. Also disappointing is the three percent contribution from the Regional Delegation of the Ministry of Environment. One would expect that these departments would do more in terms of raising awareness about waste resources, and the effects of poor waste disposal on public health and the environment in the community.

Responding to a question about the content of messages, they received from their sources of education, 35 percent of household survey respondents report that they have received messages related to hygiene and sanitation. A further 20 percent indicate that messages from health personnel call on the population to adopt proper waste management strategies to reduce the spread of malaria, cholera, and other diseases. An additional 21 percent have received information related to waste handling and the environmental impact of poor waste disposal. Another three percent have heard messages advocating waste separation. Two percent mentioned messages about pollution control and 17 percent about waste degradability, the importance of a green world and the importance of composting. Three percent of respondents have received messages addressing the risk of ozone layer depletion and the need to protect the ozone layer.

In spite of the low level of environmental education provided by the authorities, Bamenda inhabitants are generally aware that poor waste disposal has negative effects on health, natural resources, and the environment. When asked whether they could find any association between waste disposal practices and public health risks, 77 percent of the respondents attribute some common illnesses, such as malaria and typhoid fever in their communities to poor waste disposal. An additional 17 percent have mentioned cholera, cough, rashes and river blindness as common ailments. Six percent pointed to diarrhoea/dysentery. A prince of the Mankon Fondom, where the municipal waste dumpsite was located at the time of field data collection decried its location and the techniques of treatment at the site. He said the constant burning at the dumpsite produces a chocking atmosphere, causing his family members and other persons of the community to become ill periodically. He added that before the relocation of the dump site from Mile 6 Mankon to Bagmande quarter in rural Mankon, his children did not fall ill so often. In fact, he wished the dumpsite be relocated, though he had no suggestion for a new location. Caution should be taken that there is no clear proof that the illnesses mentioned by the respondents are not caused by factors other than waste. The responses do, however, serve to indicate that the people think there are effects. It is also important to note that recent information from the council reveals that the lone municipal landfill/dump site has been relocated to the out skirts of Bamenda III Sub-Divisional Council area in the rural neighbourhood of Mbelewa.

In summary, the research findings presented in this chapter have addressed the third research objective of this study that sought to analyse household waste management activities from a gender perspective. The household solid waste management activities considered included storage, sorting, collection and transportation, and disposal. These included techniques and frequency of waste transportation and the actors involved. In terms of household solid waste storage, the findings reveal that a variety of waste storage containers are generally used to temporally store waste. Plastic material containers are most used (70 percent) including bags, baskets and buckets. In terms of waste separation, it was observed that a majority of households do not sort waste before storing or disposing. In households where waste sorting is done, waste is generally separated into biodegradable and non-biodegradable waste, with the former as the dominant waste type generated. The female adult is the most active gender in the household waste separation activity.

Primary waste transportation (movement of waste from house to secondary collection point for onward disposal) is strongly gendered, with children and especially the boy child below the age of 16 and mothers as the main transporters. Waste is disposed of at varied sites, with only 32 percent of households using the conventional waste disposal sites provided by the council. This implies that a majority use unorthodox sites, resulting in outcomes that are likely to be detrimental to the environment and public health. Environmental education by the council and agents of the Ministry of Environment and Nature Protection and Sustainable Development prove to be very limited. However, people are generally aware of the harmful impact of poor waste disposal and believe it has a negative effect on public health and the environment. Unfortunately, this general level of environmental consciousness has not translated into sound waste management practices as exhibited by household waste disposal practices for neither men nor women.

Chapter Six

Market Waste Generation and Characterisation

This chapter presents research results from the nine markets that operate daily within Bamenda City Council. Findings are based on a survey of market vendors, items sold and activities carried out in the markets, which have the potential to generate waste. These findings address the fourth research objective of this study, which sought to determine the nature of market solid waste generation, the characteristics of the waste, and the nature of waste management practices, and their implications for the efficiency of Council waste collection and disposal strategies.

6.1 Infrastructural Profile of Markets

The infrastructure of the market is classified into permanent and non-permanent vending spaces with differing features that provide different advantages or risk levels. Vending spaces with permanent infrastructure constitute lock-up stores and open-slabs. The lock-up stalls (stores) with permanent infrastructure guarantee stability, provide greater chances of gaining and maintaining clientele, and offer greater security and safety of vendors and goods from damages or inconvenience caused by rain, sun and even burglary. Permanent open-slab stalls provide most of the advantages of lock-up stalls with the exception of security from theft. For these convenience and safety reasons, the rental fees for these permanent vending locations are relatively very high compared to those of non- permanent infrastructures. Non-permanent vending spaces include temporary ground space and thatched open space stacked with sticks and covered with aluminium roofing sheets, leaves and grass, palm fronds or umbrellas (Figure 6.1). The risk level in terms of security from the elements of weather and thieves is very high. Especially for those who operate in the open without a structure, the risk of losing clientele is very high, as they have no permanent location.

Plate 6.1 Temporary Ground Space
Source: Akum 2013

Table 6.1 present data showing the distribution of types of infrastructure found in the nine daily markets that operate in Bamenda City Council. Six of the nine daily markets are located in Bamenda II, two in Bamenda III and only one is in Bamenda I. According to the survey results from all nine markets, 57 percent of all vending spaces across all the nine markets are permanent structures with 57 percent of them lockable. However, there is considerable variation among different markets. For example, non-permanent infrastructure dominates in the Bamenda City Council Food Market (81 percent) unlike in Bamenda City Council Main Market, where most of the stalls (70 percent) are permanent.

6.2 Demographic Profile of Market Vendors

In this section, we present the demographic profile of market vendors in terms of sex, age and education. Table 6.2 shows that market vendors are mostly women (62 percent) against 33 percent men. The great majority of vendors are in the active working and reproductive age group, with 72 percent within the 20-39 age range. Vendors above 50 are few. This is expected as persons of this latter age group are often tired and retired and generally lack the energy required to carry out commercial activities especially away from home.

Table 6.1: Sample distribution of types of infrastructure by market by market

Sub-division	Market	Distribution of market stall by type of infrastructure									
		Non-Permanent				Permanent				Total	
		Temporary ground space		Thatched / open space		Lock-up		Open slab			
		n	%	n	%	n	%	n	%	n	%
Bda I	Bda Up-station	0	.0	0	.0	2	100	0	.0	2	100
Bda II	BCC Main Market	9	7.6	26	22.0	56	47.5	27	22.9	118	100
	BCC Food Market	62	59.4	26	23.9	15	13.8	6	5.5	109	100
	Ntarikon Market	0	.0	0	.0	27	100	0	.0	27	100
	Stadium Market	0	.0	0	.0	14	100	0	.0	14	100
	Muwachu Market	0	.0	0	.0	16	100	0	.0	16	100
	Ntatru Market	0	.0	0	.0	2	100	0	.0	2	100
Bda III	Nkwen Mile 2	2	3.7	24	44.4	28	51.9	0	.0	54	100
	Nkwen Mile 4	0	.0	0	.0	3	100	0	.0	3	100
Total		73	21	76	22.0	163	47.2	33	9.6	345	100

Source: Market Survey, 2013

Table 6.2 Sample distribution of vendors by sex, age and marital status

Sex	Frequency (n)	Percent (%)
Female	213	61.8
Male	132	38.2
Total	**345**	**100**

Age	n	%
<20	38	11.7
20-29	125	38.6
30-39	109	33.6
40-49	33	10.2
50+	19	5.9
Total	**324**	**100**

Marital Status	n	%
Married	157	50.8
Single	126	40.8
Widow/er	26	8.4
Total	**309**	**100**

Level of Education	n	%
No formal education	9	3.0
Primary	118	38.7
Secondary	129	42.3
Post-Secondary	49	16.1
Total	**305**	**100**

About half (51 percent) of the vendors in the markets are married and 41 percent are single. Widowed persons represent eight percent. In terms of education, almost all (97 percent) of vendors have had some level of formal education distributed in the following proportion; primary level (39 percent), secondary level 42 percent and post-secondary (16 Percent). With a high proportion (97 percent) of respondents literate, the information provided by them regarding waste estimates can be judged dependable. In addition, waste sensitization messages can be more easily understood and may be better appreciated where and when need be.

6.3 Distribution of Types of Market Stalls /Places by Profile of Market Vendors

6.3.1 Distribution of Type of Market Place by Sex of Vendor

Table 6.3 presents the distribution of types of market infrastructure by the sex of the vendor occupying them. Women prevail over men in shops of all form of infrastructure.

Table 6.3 Sample distribution of types of market stalls by sex

Type of market store	Sex					
	Female		Male		Total	
Non-permanent stalls	N	%	n	%	n	%
Temporary ground place	53	72.3	20	27.7	73	100
Thatched Open space	47	61.7	29	38.3	76	100
Permanent Stalls						
Lock-up	91	55.6	72	44.4	163	100
Open slab	18	55.2	15	44.8	33	100

$\chi 2$-test: $\chi 2=6.192$; df=3 P=0.103.

According to Table 6.3, women occupy 72 percent of temporary ground places, 62 percent of thatched open stalls, 56 percent of lock-up stalls and 55 percent of open slabs. Averagely, combining the results of the first and last two rows, 67 percent of shops with non-permanent infrastructure are occupied by women as against 33 percent by men. In addition, 55 percent of stores with permanent infrastructure are occupied by women as against 45 percent by men. These results indicate on the one hand that the proportion of women who occupy stalls with non-permanent infrastructure doubles those of men. On the other hand, men despite their low representation are more prominent in the occupation of shops with permanent

137

infrastructures (45 percent). This could be because men have more potential to acquire the capital required to rent this type of stalls, which are relatively more expensive than women are. It can also be because non-permanent infrastructure is an indication of inferior status, from which men shun away.

6.3.2 Distribution of Type of Market Place by Age of Vendor

Overall, persons within the age range 20-39 dominate, occupying 72 percent of all vending places (Table 6.4). They occupy 75 percent of temporal ground places, 61 percent of thatched open space, 77 percent of lock-up stalls and 67 percent of open slabs. Averages of the vertical summation for the different types of stores reveal that 82 percent and 18 percent of vendors below the age of 20 take up non-permanent and permanent infrastructures respectively. Those in the age group of 20-29 take up 54 percent and 46 percent; 30-39 occupy 36 percent and 64 percent; 40-49 take up 30 percent and 70 percent; and vendors above 50 occupy 37 percent and 63 percent of non-permanent and permanent infrastructure respectively.

Table 6.4 indicates that 80 percent of vendors occupying temporal ground places in the markets are mostly persons below the age of 29. This can partly be explained by the fact that persons of this age are still financially wanting and undecided in career choices. This indecision is a common phenomenon particularly in communities of the developing world with a high rate of unemployment. Most vendors (42 percent) above the age of 40 occupy stalls of permanent infrastructure. It may be that most persons of such age have already gained some financial stability and established a business career and can afford to pay the higher amounts required or that those who were unsuccessful dropped out of selling in the market before reaching this age.

6.3.3 Distribution of Type of Market Place by Marital Status of Vendor

Table 6.5 shows that 51 percent of vending places are occupied by married vendors as compared to 41 percent by unmarried. The only exception is temporal ground spaces, where unmarried persons occupy only 46 percent.

Table 6.4 Sample distribution by type of market stores by age

Type of market store	Age											
	<20		20-29		30-39		40-49		50+		Total	
Non-Permanent	n	%	n	%	n	%	n	%	n	%	n	%
Temporary ground space	20	23.5	48	56.5	16	18.8	0	.0	1	1.2	85	100
Open space	11	15.9	19	27.5	23	33.3	10	14.5	6	8.7	69	100
Permanent												
Lock up	3	2.2	48	35.0	58	42.3	18	13.1	10	7.3	137	100
Open slab	4	12.1	10	30.3	12	36.4	5	15.2	2	6.1	33	100
Total	38	11.7	125	38.6	109	33.6	33	10.2	19	5.9	324	100

Source: Market Survey, 2013

Table 6.5 Sample distribution by types of market stalls and by marital status

Type of market stalls	Marital status							
	Married		Single		Widow(er)		Total	
Non-Permanent Stalls	n	%	n	%	n	%	n	%
Temporal ground space	40	46.5	43	50.0	3	3.5	86	100
Open space	31	47.7	27	41.5	7	10.5	65	100
Permanent Stalls								
Lock up	71	55.9	46	36.2	10	7.9	127	100
Open slab	15	48.4	10	32.3	6	19.4	31	100
Total	157	50.8	126	40.8	26	8.4	309	100

χ2-test: χ2=11.486; df=6 P=0.074.   *Source: Market Survey, 2013*

Findings as shown by data in Table 6.5 suggest that the married vendors may have greater financial resources, perhaps due to support from their spouses. A low percentage of stalls are occupied by widows. However, the relationship between type of market space and marital status of the vendor is not statistically significant (χ2-test: χ2=11.486; df=6 P=0.074).

6.3.4 Distribution of Type of Market Place by Level of Education of Vendor

Table 6.6 presents research results indicating the infrastructure of market places and the level of education of vendors occupying them.

Table 6.6 Type of market stall and education of vendor

Type of market store	Level of education									
	No formal education		Primary		Secondary		Post-secondary		Total	
Non-Permanent Stalls	n	%	N	%	n	%	n	%	n	%
Temporal ground space	3	3.5	32	37.6	30	35.3	20	23.5	85	100
Open Thatched space	1	1.6	28	44.4	23	36.5	11	17.5	63	100
Permanent Stalls										
Lock up	3	2.4	46	36.2	60	47.2	18	14.2	127	100
Open slab	2	6.7	12	40.8	16	53.3	0	.0	30	100
Total	9	3.0	118	38.7	129	42.3	49	16.1	305	100

Source: Market Survey, 2013

The survey results presented in Table 6.6 further demonstrate that persons with primary and secondary education dominate in all categories of stalls. Combining results for the first two rows, an average of 41 percent of vendors who occupy non-permanent and 21 percent have attained secondary education status. Another 21 percent have post-secondary level of education. Conversely, permanent stalls are more likely to be occupied by vendors with at least secondary level of education (50 percent). This suggests again that those in the permanent stalls may probably have more financial resources. Strangely enough, more vendors with post-secondary level of education occupy more non-permanent than permanent structures. For example, they occupy 24 percent of all temporary ground spaces and 18 percent of all open thatched spaces as opposed

140

to 14 percent of lock-up stalls and zero percent of open slab structures.

6.4 Waste Generation Propensity

The amount and type of waste a market generates is generally linked to the type of products sold. In addition are the complementary activities vendors carry out during the time they spend in the market. Some of them include peeling oranges, shredding egusi (cereal), selecting vegetables, slicing eru (vegetable with climbing stems), etc.

Plate 6.2: Public waste site at the Bamenda City Council Food Market

Waste is also produced from the material used as packaging for items that enter or leave the market site. For example, items such as tomatoes, yams, cocoyam and spices are normally brought into the market in baskets made of bamboo, palm material, and covered with grass or leaves. In the same way, transformed products like soap, cosmetics and others are brought into the market in cartoons by wholesalers. The presence of a wide variety of products, processed and un-processed, the different activities such as milling, hair dressing and catering imply the generation of waste of multiple types.

Plate 6.2 shows a public waste site at the Bamenda City Council Food Market comprising of biodegradable and non-biodegradable waste materials. The biodegradable waste materials include grass, wooden baskets and discarded vegetables, and non-biodegradable waste materials consist of plastics, cartoons, rubber, etc.

6.4.1 Types of Items/Services

Table 6.7 reports the kinds of merchandise sold and services rendered in each type of vending space. The food items sold generally include vegetables (huckleberry, bitter-leaves, cabbages and other greens), fruits (oranges, paw-paw, mangoes, pears, guavas, pineapple, etc.), legumes (beans), cereals (corns), groundnuts, tree food crops (plantain and bananas), spices, canned/packaged foods, tubers (cassava, cocoyam, potatoes, etc.), oil (palm and vegetable), fish/meat, poultry products, etc. The non-food items most commonly available include textile, clothing and footwear. Also sold are household utensils, cosmetics, jewellery, books, automobile spare parts, etc. Activities and services are dominated by sewing and embroidery, eating/drinking places, mending/repairs, milling, hairdressing, and communication (call box). Overall, unprocessed food items are sold in 39 percent of all market spaces and processed and other non-food items in 43 percent as presented in Table 6.7. Services are provided in 18 percent. There is a clear relationship between the type of vending space and the types of merchandise or services provided by its occupants. Unprocessed food items are mostly sold in non-permanent spaces.

Averaging results for the first two rows (temporal ground space and open space), Table 6.7 indicates that unprocessed foods are sold in 53 percent of non-permanent structures, processed foods and non-food items occupy 35 percent and services 12 percent. Considering the type of items sold in stores with permanent infrastructure (lock-up and open slabs), general trend of results indicate the following: Unprocessed food items are found in 41 percent of the stores, another 41 percent of permanent stores are occupied by vendors who sell processed and non-food items and 18 percent by services. Unprocessed food items tend to occupy mostly the temporary ground space, open thatched spaces and open slabs to maximize the visibility of the merchandise. This is much needed because most of the goods are perishable and need to be sold as soon as possible. Some of the goods need a great deal of space by nature, for example bunches of plantain, baskets of tomatoes, bags of potatoes and yams, etc. They also need to be in the open to reduce heat that might hasten the rate of perishability. In order to avoid the effect of direct sun light, some of the products are usually covered or placed under some temporal shade. Figure 6.1 indicate the different markets and the main types of items sold.

Table 6.7 Sample distribution by item sold and by type of stores

Type of market stall	Item sold/Services							
	Unprocessed food items		Processed food & non-food items		Activities/ Services		Total	
Non-Permanent	n	%	N	%	N	%	n	%
Temporary ground space	55	64.0	28	32.6	3	3.5	86	100
Open space	31	41.9	28	37.8	15	20.3	74	100
Permanent								
Lock up	22	15.5	81	57.0	39	27.7	142	100
Open slab	21	65.6	8	25.0	3	9.4	32	100
Total	129	38.6	145	43.4	60	18.0	334	100

χ^2-test: $\chi^2=70.341$; df=6 P<0.001   *Source: Market Survey, 2013*

We observe from Figure 6.1 that while almost all types of items are sold in all the markets, there are a few peculiarities. The Mankon-Bamenda City Council Market, which is the central and largest market, has a good representation of all items and services. In line with its name, the Bamenda City Council Food Market is dominated by the sale of food items, while the Stadium Market features automobile spare parts and household utensils. The Nkwen Mile 4, the Ntatru, Bamenda Up-Station and Mankon Mile 8 markets are identified with catering services, other food items and general provisions, as they are situated in motor parks used mostly by travellers. The Nkwen Mile 2 and Ntarikon Markets are of average size and display coverage of food and non-food items and some services.

143

Figure 6.1: Distributions of markets and items sold

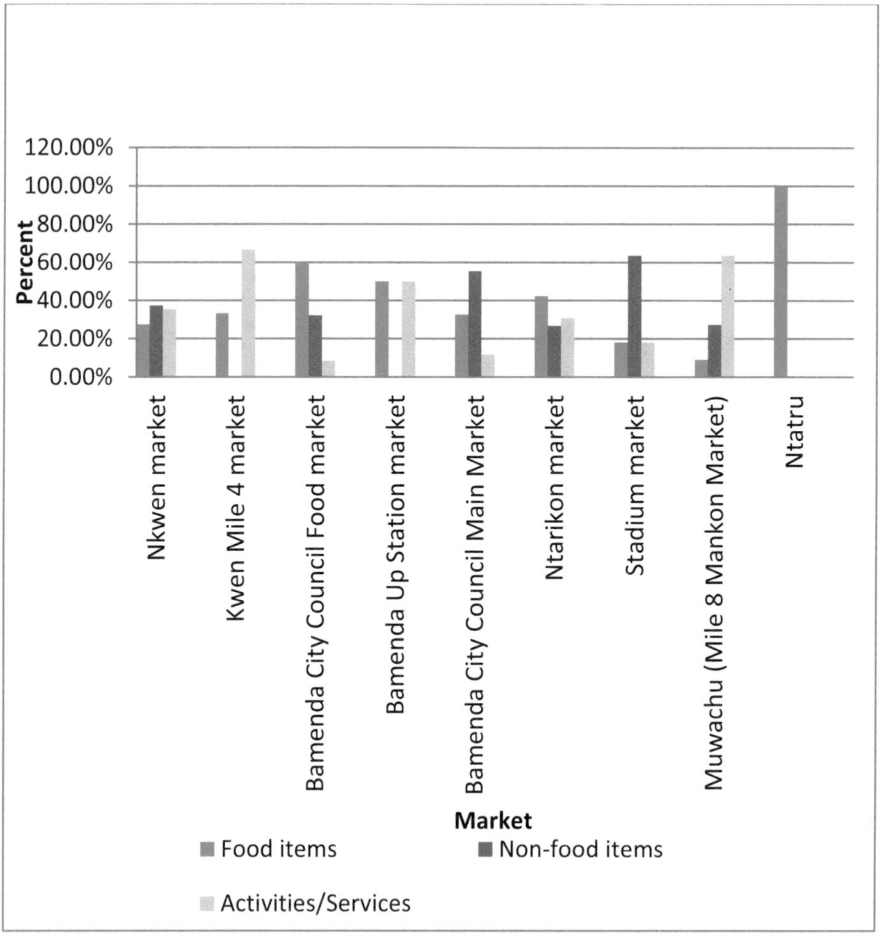

Source: Market Survey, 2013

6.4.2 Relationship between Sex of Vendor and Items Sold

Table 6.8 shows that the types of products sold and services provided at the market place have a strong and significant relationship with the sex of the vendor.

Table 6.8 Sample distribution by sex and item sold

Sex	Item sold/Activities							
	Food items		Processed food and non-food items		Activities/Services		Total	
	n	%	n	%	N	%	n	%
Female	119	55.7	51	24.1	43	20.3	213	100
Male	22	17.0	87	66.0	23	17.0	132	100
Total	141	41.3	138	39.7	66	19.0	345	100

χ2-test: χ2=47.766; d.f.=2; P<0.001.   *Source: Market Survey, 2013*

Fifty-six percent of the female vendors are involved in the sale of unprocessed food items, 24 percent in processed and non-food items and 20 percent in services such as catering. A majority of the male vendors (66 percent) sell non-food items, such as clothing and kitchen utensils, while 17 percent sell food and another 17 percent provide services. The high representation of women in the sale of food may be due to the fact that such businesses do not initially require heavy capital, as food items are mostly sold in temporary open ground space that require a daily toll of only 100FCFA, much lower than a lock-up store. The results can also be interpreted to mean that there is a tendency for women to choose activities and items that are closely related with their household reproductive activities like food preparation. Male dominance in selling non-food items may be explained by the fact that they are likely more able to afford the investment in merchandise required for such businesses. This sex and item related results prompts one to suggest that women tend to generate most of the putrescible waste given their close involvement in the sale of unprocessed food products (Plate 6.1). On the other hand, men will contribute more to the production of non-biodegradable material as plastics and cartons from which they unpack textiles, footwear and kitchen utensils.

6.4.3 Types of Packaging

Packaging contributes to the type of waste the Council and other waste operators have to handle, as the packaging usually becomes waste after sale of the products.

Figure 6.2: Types of material used for packaging

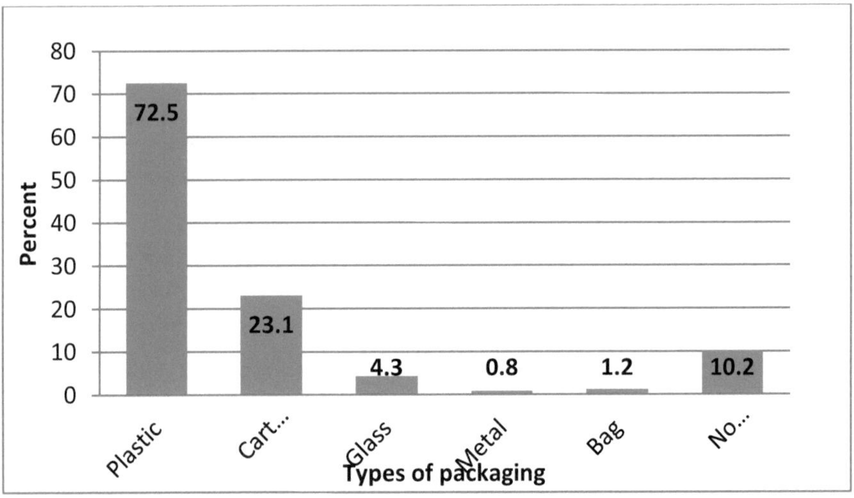

Source: Market Survey, 2013

Packaging used when products are brought to the market and packaging provided at the time of sale of products (depending on the type of item) all contribute to market waste generation. Retail activities in the markets contribute to waste generation as cartons, bags, and baskets from wholesalers are separated and sold individually into smaller units. According to survey results Figure 6.2, 73 percent of material used for packaging is plastic followed by carton/paper (23 percent).

Information from vendors indicating whether the items they sell require them to be packaged reveals the following results. In general, ten percent of items are sold without packaging provided for by the vendors. Referring to unprocessed food items, 50 percent of the vendors indicate that the items they sell are not often accompanied by packaging. Often, the buyers come along with containers to collect items as cocoyam, potatoes cabbages, etc. Some items as plantain are often carried away in bunches. The situation is different with unprocessed and non-food items. Results indicate that only four percent of processed and non-food items are sold without packaging. Bearing in mind that unprocessed food items and non-food items constitute 41 percent (Table 6.7) of items sold in the market, and that 73 percent (Figure 6.2) of packaging material used are plastics, we can infer that non-food items sold in the market possess a high propensity to contribute non-biodegradable material into the municipal waste stream.

6.4.4 Waste Generation Quantities

The volume of waste generated varies with vendor and market. Table 6.8 present results of waste amounts generated in the different markets and measured in litres as estimated by the vendors. The survey data reveal that the average quantity of waste generated daily by a vendor is slightly over half a sac & motto bag (12 litres i.e. half the size of the bag), the exact average at the Nkwen Mile 2, Ntarikon and Stadium markets. The Bamenda City Council Food Market and the Muwachu Market have slightly higher amounts of 13 litres each. The Bamenda City Council Main market has slightly below the average (9 litres). The average for the Bamenda Up-Station is very far below the average (1 litre). This could be considered an outlier and partly explained by the fact that Bamenda I is a generally lowly populated area.

The differences in the waste amounts generated per capita are unrelated to market size. For instance, the average amount of waste generated per vendor per day in the Bamenda City Council Main Market and Bamenda City Council Food Market, which are considered to be of higher status both in terms of size (area coverage) and products sold is about the same as in the Ntarikon, Stadium, Muwachu markets. This can partly be explained by the fact that activities and the nature of the product sold have a greater influence on on-site waste generation. The relatively smaller markets of Muwachu, Ntarikon and Nkwen Mile 4 markets serve as travel stations, where food and instant like fast foods and fruits are prepared for instant consumption. Such activities and products have relatively high waste generation potential from peelings and packaging of products.

Table 6.9 Waste generation volumes in litres

Market	N	Mean	Minimum	Maximum	Std. Deviation
Nkwen market	13	12	5	40	9
Bda City Council Food market	93	13	4	60	8
Bda Up Station market	1	1	1	1	
Bada City Council Main Market	31	9	4	20	3.8
Ntarikon market	6	12	5	20	6.6
Stadium market	8	12	10	20	4.6
Muwachu (Mile 8 Mankon Market)	3	13	10	20	5.8
Ntatru	1	10	10	10	
Total	156	12	1	60	7.4

Source: Market Survey, 2013

6.5 Market Waste Management

6.5.1 Perception and Types of Waste

How market vendors perceive waste can affect their attitude to waste generation activities and effect on waste minimisation. It can also determine vendors' waste management practices that can either promote better practices and enhance municipal waste management efforts or worsen the situation. From research results, market vendors define waste variously. A large majority (71 percent) perceives waste as anything that is not useful (useless), while 15 percent consider waste as anything intended for disposal and seven percent as pollutants (whatever threatens the environment and human health). A very small percent of vendors see waste as recyclable material (one percent). The ways vendors perceive waste reflects that of household respondents as presented in chapter four (see section 4.3.1).

From observation, waste generated at the market sites can be categorized into biodegradable and non-biodegradable material. Waste at the market place results from sweeping and cleaning of vending places, eating, packaging from retailing activities as whole cartons and baskets or bags are unpacked, fall-offs from items such as vegetables, plantains, tomatoes, and onions. Some of the waste is also a result of activities that the vendors carry out during their stay in the market either for pleasure or out of necessity. For most women, the pastime activities are an extension of their household gender reproductive roles. These include shredding egusi, groundnuts and other nuts, sorting vegetables, peeling oranges and other fruits. These activities are meant to prepare items for sale or to facilitate their task of food preparation when the vendors get home. These tasks undertaken by the market vendors attest to the utility of the gender analytical tool that describes the involvement of participants in activities by gender of the actor. It is observed that men rarely carry out such pastime activities that generate waste. They rather engage in discussions or games of cards while at the market place.

There is also a temporal element in the type and amount of waste that markets generate. Observation and interviews with market authorities and vendors reveal that on Fridays, Saturdays and Sundays, markets tend to generate high amounts of waste, especially organic waste. These are busy days for vendors and buyers. This result corroborates that of household on-site waste measurements, which reveals that greater amounts of waste are generated during weekends that mid-week. Generally, more waste is generated in the rainy season, both in quantity and variety, than the dry season. This is the period when vegetables, fruits and other food items ripen. During the rainy season, insects like flies, mosquitoes and other insects are also very common and can transmit diseases as they feed on fruits as mangoes.

6.5.2 Waste Storage

Market vendors store waste in containers made of different materials. Plastic based containers (87 percent) are primarily used to store waste at the market. Plastic storage vessels include plastic bags, buckets and baskets. The specific material out of which waste storage containers are made and the size of the containers determines their potential to become waste products in themselves and their capacity to hold waste. The frequency with which vendors dispose their waste may relate to the size of the waste storage container. Wood-based storage vessels represent only 12 percent, mainly in the form of baskets and cartons initially used to package goods brought into the market by whole sellers. Metal is the least used material, with a representation of one percent, usually in the form of wheelbarrows.

Survey results suggest that the length of time vendors store their waste also depends on the amount of waste, type of waste, the ability to pay someone to carry waste to bin and the proximity of the nearest waste disposal facility. A majority of vendors, 74 percent, store their waste for a maximum of one day, and 87 percent store their waste for three days or less. Reasons advanced by those who store their waste for less than three days include the fact the disposal bins are close, hygienic reasons, and the desire to avoid flies and smell. Those who accumulate waste for more than three days offer reasons such as lack of money to pay someone to transport waste or that their businesses generate very little waste. Comparing the length of time market vendors store waste with that of households, it is observed that the former do it for a relatively shorter duration. This can partly be explained by the proximity of authorized waste bins to market vendors.

6.5.3 Waste Separation

In response to the survey question demanding whether or not market vendors separate their waste, results reveal that a large majority of vendors (82 percent) like households do not separate their waste. Such actions have important implications for waste recycling and treatment efforts. When asked about their reasons for not separating waste, 49 percent of respondents simply say that they perceive waste as waste and therefore useless and not in need of separation. Seventeen percent indicate they generate only a single waste type, and 25 percent consider time and labour required as the obstacle. A few say that they do not like to deal with more than one waste container, and others added that they lack the space and containers needed (three percent). The rest (five percent) cite Council's apparent lack of interest in waste separation, as there is only one bin with no provision for waste separation and the Council does not demand that waste be separated. The 59 respondents who say that they do separate waste advanced the following reasons: their wish to separate plastic and non-biodegradable waste (17 percent) and the desires to use organic waste as manure (19 percent), provide food for domestic animals (47 percent), offer for no pay to those who request for it (three percent) and sell for money (14 percent). Some waste material sold include cartons and bags from which products like tablet soup and rice for retail are gotten, and peelings of fruits as paw-paw and pineapples, and left-over food material from restaurants used as animal feed.

Waste awareness through education is at a very low level for market vendors. When asked whether they had ever had any education on waste issues, only five percent of vendors answer yes. Those who answer yes identify the Council as the source of awareness and the message is to keep their environment clean and dispose of waste in the bins. It is striking that only 18 vendors (5.7%) out of 318 in a public place like the market had been informed about waste issues by the Council. Could it be the strategy used is inadequate? Strangely enough, interviews with officials contradict the survey results. For instance, the market manager of the Bamenda City Council Food Market, Madam Mambo, suggests that vendors are instructed to clean the vending spaces at the close of each day. As a follow up, sanitary inspectors do daily controls in the mornings to identify defaulters and impose a fine of 10.000FCFA ($20). This amount, some vendors reported, is a major deterrent and reason why they make every effort to keep their spaces clean.

6.5.4 Waste Disposal

Market vendors suggest that there are multiple sites and ways by which they dispose of waste at the market place. Table 6.10 present these and methods. Results show that the great majority of market vendors (83 percent) dispose of their waste in the communal waste-bin provided by the Council. Some vendors take waste home or use it as feed for domestic animals, others offer to those who have use for it, and some sell it. Only four vendors reported recycling waste (Table. 6.10).

Table 6.10 Distribution of respondents by disposal sites

Disposal site	Frequency (n)	Percent (%)
Throw in public bin	276	82.5
Burn at disposal site	6	1.8
Take home	16	4.9
Feed for domestic animals	19	5.8
Sell	11	3.4
Give to those who need it	10	3.1
Recycle	4	1.2
Throw in nearby gutter	1	0.3
Total	343	105.2

Multiple responses

When respondents were asked whether they had problems handling their waste, only 17 percent said yes. Some who said they had difficulties cite inability to pay someone to carry their waste to disposal sites. Others mentioned the irregularity of waste collection by Council or lack Council provided bins. It is observed that most of the vendors with these problems have their vending places located more than 100m from the authorized waste disposal site, which they consider very far away. Survey, interviews and observation carried out in April, 2013 indicated that there was no public trash bin in some of the markets, namely Muwachu (Mile 8 Mankon), Stadium and Mile 4 Nkwen markets. In the Nkwen Mile 4 Market, for instance, all vendors after sweeping carry their waste to the nearby stream where a big open dump exists. Nkwen Market as at time of field survey in 2013 had two public dustbins, Bamenda Up-Station one and Ntarikon Market two. In the Bamenda City Council Food Market, there were only three bins, which some vendors said were too few for such a large market. This respond is further testified by the overflow at the authorized public waste disposal site as indicated on Plate 6.2. Contradictory responses were obtained from market managers and vendors as to the frequency of waste collection by the Council. According to the market authorities, waste is collected at least once every two days or on a daily basis. Vendors, however, indicated that the waste collection team visits once a week and sometimes once every two weeks. We observed that there are always piles of waste in and around the trashcans.

Summarily, this chapter has presented findings addressing research objective four that sought to examine market waste generation and the characteristics of waste. In this regard, the study looked at the different markets, types of products sold, services and activities of vendors and their propensity to generate waste. Research findings reveal that there are nine markets within the City Council distributed in the three subdivisions that operate on a daily basis. Of this number, there is one in Bamenda I, six in Bamenda II and two in Bamenda III. Vending places in the market are characterised as either permanent or non-permanent depending on the nature of the infrastructure. Non-permanent infrastructure comprises vending places or stalls, which are described as temporal ground spaces and thatched open spaces. This category of vending spaces is generally less expensive to acquire but are equally the most insecure. The second category of vending spaces is those made of permanent infrastructure as bricks and cement. Shops with such infrastructure as described as lock-up and open slabs. They are relatively more expensive to acquire and are better secured.

Considering the demographic profile of market vendors within the Bamenda City Council, findings disclose the following. Market vendors are mostly women and are within the active working and reproductive age range of 20-39. Slightly above half (51 percent) of all market vendors are married. Almost all (97 percent) of vendors are literate with a majority (81 percent) with either primary or secondary levels of education. Looking at the distribution of types of market infrastructure by demographic profile of market vendors, findings reveal that women prevail over men in the occupation of shops of all form of infrastructure. However, more men occupy shops with permanent structures than temporary ones. Age wise, persons within the age range 20-39, occupy 72 percent of all vending places.

Focusing on the sale of the different types of market products and services that have the potential to generate on-site waste, women again dominate and can therefore be considered as main actors in market organic waste generation. On the other hand, men are most involved in the sale of non-food products that create waste via unpacking of goods from wholesale packaging for retail. As such, men like women in such businesses, tend to generate inorganic waste materials. Packaging used to parcel goods for customers, which is predominantly plastics, has the potential to contribute to the municipal waste stream. Consequently, packaging material used in the market has implications on the type of waste generated in the household. This is real as the high proportion of plastic bags used for parcelling is strongly reflected in the non-biodegradable component of household waste. Plastic material waste storage containers are equally the most used. A majority of vendors do not separate their waste. Mass education for environmental awareness is not a common practice. Consequently, disposal practices by vendors are both orthodox and unorthodox, with the use of communal bins more common than is the situation with households that use unauthorized sites more.

Chapter Seven

Assessment of Council Waste Collection Strategies

In this chapter, findings related to objective five of this study are presented. This objective sought to determine the extent to which municipal solid waste collection policies and strategies influence households' and market vendors' waste generation and handling practices and their implications for effective management. The goal is to determine the extent to which the Council is fulfilling its obligation to maintain city hygiene and sanitation through its waste collection, transportation and disposal strategies. Included in this chapter are findings concerning the extent to which municipal waste policy and strategies are gender sensitive. The goal is to make evident the need to mainstream gender in municipal waste management strategies in order to enhance efficiency.

7.1 Council Waste Policies

Information from documents, particularly presidential and ministerial decrees, texts, order and service notes, reveals that governance and participation in municipal solid waste management responsibilities in Cameroon, and certainly in Bamenda, is the concern of different government departments including: the Ministries of Commerce, Economy and Finance, Health, Urban Development, Town Planning and Housing, Environment, Forest and Nature Protection and Sustainable Development, Mines, Water and Energy and Territorial Administration and Decentralisation (MINATD). The state makes the laws and regulations and defines the responsibilities of the different stakeholders or actors.

For example, the Ministry of Urban Affairs, according to Decree No. 98/153 of 24 July, Articles 22-25, assign the Ministry with the responsibility to ensure general cleanliness and drainage, solid waste management, hygiene and sanitation of the cities. The Ministry of Environment, Forest, Nature Protection and Sustainable Development determines national environmental policy plans, with a further task to coordinate and ensure the implementation of the policies. Proof of this is the national environmental law no 96/12 of August 1996. Articles 42 to 53 of this law focus on municipal solid waste management practices for Cameroon. According to this 1996 law, and Law No. 2004-18 of 22 July 2004 defining the duties and powers of a mayor at the head of a city council, municipal solid waste management in Cameroon cities, including waste collection, transportation and disposal, is the statutory responsibility of the Council (Law No. 96/12, 1996).

The Ministry of Territorial Administration and Decentralisation (MINATD), under whose jurisdiction councils are found, is charged with the responsibility of ensuring good sanitation and waste management among other duties (Decree No. 98/147 of 17 July 1998). Specifically, Article 24 assigns the Ministry the responsibilities of elaborating plans for evacuation and treatment of solid waste, carrying out research on improving collection and transportation, supervision and coordination of collection and transportation, and educating the public about the practice of pre-collection of MSW. Finally, the law on decentralisation in Cameroon (Bill N0 762/PJL/AN on the Orientation of Decentralization No 51/AN, June 2004) devolves competences and resources from centralised units to peripheral authorities. In line with this law, the Ministry of Territorial Administration and Decentralisation devolves powers to regions, city and sub-divisional councils whom together can bring meaningful change to the disturbing waste management situation in our cities. Unfortunately, information from interviews and secondary sources reveal that administrative bottlenecks, as well as conflicts in roles and authority, tend to slow down the waste management process (Manga, 2007; Achankeng, 2005). Effective coordination and collaboration between the different stakeholders of influence and importance (often lacking) is required to obtain maximum results.

Among the so many duties accorded to a city council are the following ones related to waste management issues and maintaining sustainable, clean urban environments. These are: creation, maintenance and management of greens, parks and gardens; management of city lakes and rivers; monitoring and control of the management of industrial refuse; cleaning of city roads and areas;

collection, removal and treatment of household waste; creation, development, maintenance, and preparation of urban environmental action plans, especially regarding the fight against pollution and other forms of nuisance, protection of lawns, amongst others (Law No. 2004-18 of 22 July 2004, Section, 110). Bamenda is a city council to which competences and corresponding appropriate resources have been devolved to from the Ministry of Territorial Administration and Decentralisation (MINATD). It has power and obligation to set local policies, strategies and practices through its technical services to realise duties assigned to it by the law. Waste management is one of such responsibilities. Figure 7.1 presents the waste management unit in the administrative flow chart of the Bamenda City Council.

Figure 7.1: The Waste Management Unit in the Bamenda City Council administrative Structure

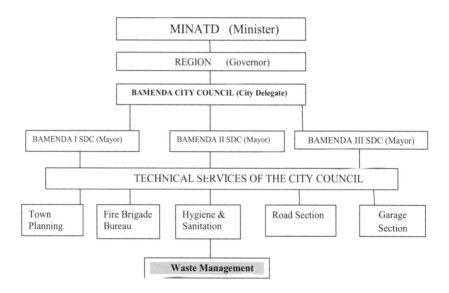

7.2 Council Waste Collection Strategies

The strategies used by the council to collect and transport municipal waste strongly influence the waste management practices of households and market vendors. I illustrate this principle by focusing on the equipment used by the City Council the personnel involved in and the organization of waste collection and transportation. It is important to note that, though three sub-divisional councils (communes) exist, the city council has not devolved to the mayors of these councils the powers of managing waste in their various administrative units. These communes however report to the City Council indicating need for intervention.

7.2.1 Municipal Waste Management Equipment in the BCC

The Council use diverse types of equipment to collect and evacuate waste. According to an interview with the Chief of Service for Hygiene and Sanitation, the pool of equipment at the disposal of the Council is grossly insufficient to effectively collect waste. It includes 4 rear-loader compactor trucks, 1 front-end loader (in bad condition), 4 can carriers, 37 communal (public) bins and a 7 ton tip-up truck. In addition, a bulldozer truck is hired from the roads department when the need arises to cover waste at the open-dump site with earth. The rear-loader compactor truck collects waste from the communal bins. It does so by hooking to the sides and lifting the container while emptying its content into the truck's waste compartment. The service of the truck is faster than that of manual labourers, but it is sometime slowed by the fact that there is usually waste dumped outside the bin, requiring workers to sweep and transfer it into the truck with brooms and shovels. The front-end loader often transfers waste especially that dumped on the ground along roadsides, which is common practice, into the tip-up truck for onward transport to disposal site. Interview with the Chief of Service of Hygiene and Sanitation at the City Council reveals that communal bins are placed at strategic locations within the City. The locations are chosen by the council along the tarred roads and some on all-season motorable earth roads based on the distribution of households and accessibility to council waste collection truck for evacuation. The size of the bin is 1.000 litres; some are made of plastic and others of metal.

7.2.2 Composition of the Municipal Waste Management Team

The personnel charged with municipal waste management include staff directly employed by the City Council and others from

158

contracted companies. The personnel of the municipal waste management department at the City Council constitute the Chief of Service in charge of administration and supervision, a secretary and the technical personnel. The technical personnel are charged with repairs and maintenance of vehicles and communal trashcans. Included are members of staff of the waste collection team. The team is made up of the driver of the truck and the labourers. The labourers assist in directing the driver and in gathering and transferring spilt-over waste that is picked up by the rear and front loaders from the ground and loaded into the trucks. These assistants use manual tools as brooms, spades and rakes.

In addition, cleaners from private waste operators have remunerated contracts with the Bamenda City Council. They sweep the streets and clear the gutters, placing the waste at accessible places for the council waste collection team to pick up. Some of these workers also assist in transferring waste into the trucks at the secondary collection points and from the trucks at the open dumpsites. According to the head of the department of the hygiene and sanitation, contracts have been signed with four private operators to do this work. They are allocated to the different subdivisions as follows: NDIFORSON & SONS for Bamenda I, SIMBA Plc, and LOW PROFILE Enterprise for Bamenda II, and ENTECO Enterprise for Bamenda III. Bamenda II requires the services of two contractors because of its larger area and high population, which generates more waste. Each of these private waste operators provides 24 workers for the tasks described above. The number of cleaners is further increased during the long holiday period (June to August), as council recruits students on holiday in a bit to provide temporary employment for them. The interviewee (chief of service) conceded that, despite all efforts at providing waste infrastructure, some parts of the town are not provided with council waste collection services, and in most parts, service provision is deficient. We discuss reasons advanced for this limited provision in subsequent sections.

7.2.3 Mobile Van Waste Collection Technique

The mobile van waste collection technique is a popular method used by the council to collect waste in the municipality. By this method, an itinerant waste pick-up truck goes into the streets and household members run to it to deposit their waste as it announces arrival (Plate 7.1).

Plate 7.1: Mobile Van Waste Collection   Source, Akum, 2015

This method is highly desired by the population, unfortunately, the service is not reliable because of the system irregular and unknown schedules. Moreover, service provision is limited to inhabitants who live along very accessible streets. These are mostly tarred roads as that from Mile 4 to Hospital Round-About. With this technique, the waste collection team makes stops to collect waste directly from homes and business premises in the course of circuiting the city. The driver hoots before arrival to alert the residents. Persons rush to the main road carrying their waste, which is deposited into the waste van with the assistance of the workers for onward transport to the disposal site. An interview with one of the council truck drivers indicates that this method of waste collection is more frequent on paved than unpaved roads. Such information complements some responses from the resident questionnaire, which indicated that the mobile van comes only once a week, once a month or, in some areas, never. With such infrequent visits, waste accumulation is high and illegal disposal is to be expected.

7.2.4 Stationary Communal Dust Bin Waste Collection Technique

The stationary communal dustbin collection technique is also much used. The limitation with this service is that the bins are inadequate in number and distant from most households. According to interviews with officials, 37 communal bins are positioned at strategic and accessible locations along the main roads. The average distance between one bin site and another is about 750m in Bamenda II and even further in Bamenda I and Bamenda III. Consequently, most people live further than 100m away from the communal waste bin location. One hundred metres is significant because it is the average distance that most households indicate their willingness to travel in order to dispose of their waste. Therefore, farther distances only imply more disposal of waste at unauthorized sites.

The waste collection team is supposed to make a daily circuit of bins starting from the council premises (recently relocated from Nghomgham to Mulang) through the city, to the landfill and back to council. According to Mr. Hilary, a member of the waste collection team, a compactor truck has the capacity to collect and contain waste from 40 bins. This implies that a single compactor truck should do a single circuit a day given that only 37 bins are available in the town. This number is however, far from the reality because additional circuits are needed because the bins are always overfilled and there are many dumpsites along the streets.

7.3 Assessing Council Waste Collection Strategies

Council's waste collection strategies clearly do not adequately meet household waste disposal needs. To determine the extent to which council waste collection strategies fall short, the following factors are considered: household proximity to a council waste collection facility; regularity of waste collection by the council; condition of the waste collection sites; collaboration between council and other waste actors; attitude of the waste collection team towards the public in the discharge of their duties; and adaptability of technology used to local environment and users of facility.

7.3.1 Location of Communal Bins and Effects on Household Waste Disposal Practices

The distance residents have to travel to dispose of their waste in the authorized communal waste collection site tend to influence their waste disposal practices. Table 7.1, based on the household survey shows the distance (estimated in the distance between electric poles) that households must cover to dispose their waste at an authorized waste collection site (stationary and pick-up van).The distance between one electric pole and another is approximately 50m.

Table 7.1 Distance between household and council waste collection site

Distance	Approximate Distance in Metres	Number (%)	Percentage
About one electric pole	50	78	26.1
About two poles	100	45	15.1
About three poles	150	45	15.1
About four poles	200	38	12.7
About five poles	250	26	8.7
More than five poles	More than 250	67	22.4
Total		299	100
Missing		40	

Household survey, 2013

Forty-one percent of households are located within 100 metres (2 electric poles) of a communal waste bin or mobile collection van and 59 percent beyond 100 metres. About 15 percent live 150 metres away and another 26 percent above 200 metres. Assuming that the distance from the households can either deter or encourage households to effectively use or not to use authorised or unauthorised waste sites, such information was solicited from household respondents. To this request, 60 percent of the respondents gave affirmative response while 40 percent said it had no effect on their use of authorized council sites. This implies that majority of households are unlikely to use the legitimate waste disposal facility because they are far off and will probably dispose of their waste at illegal sites. To determine the extent to which citizens believed that distance influences use of authorized public waste facility. Overall, 76 percent of respondents indicate that distance very much affects them, another 20 percent said not much and 04 percent suggest that it has no effect.

Distance to authorized waste collection affect households differently with varied consequences. The 92 percent of respondents, who say the location negatively affects their use of the authorized waste collection facility, attribute it to the distance being too great. As alternative measures, most of the households tend to use illegal sites such as the stream and gutters, which are closer. Below are some responses from the code quotation grounding report indicating the perceived negative consequences of distance on households by respondents:

"The distance has made me to use the stream nearby" [Female teacher, post-secondary, aged 36-40]

"It causes me to constantly burn" [Retired male, aged above 50]

"Distance discourages" [Female teacher, post-secondary, aged 31-35]

"Because I am throwing waste in the bush" [Female housewife, secondary education, aged 26-30]

"It causes us to use the stream" [Business operator, aged 41-45]

"Because it is far from my house, causing me to throw waste in the stream" [Female traditional doctor, primary, aged 41-45]

"Far away, it makes me to keep dirt longer than I want" [male trader, primary, aged 31-35]

"Far away, it makes me to keep dirt longer than I want" [male trader, primary, aged 31-35]

The remaining eight percent of respondents who suggest that location of waste site does not affect them point proximity to the sites (four percent) and one percent say the distance provide them with an opportunity to do physical exercise. Another three percent suggest it has no effect because they use their cars to dispose of the waste on their way to other functions.

Further detail from survey results reveals that, there is a statistically significant difference ($\chi2= 20.070$; df=1; P<0.001) between the percentage of males (45 percent) and female (71 percent) who say that distance affects their use of council authorized waste facility. This difference may be explained by the fact that adult males rarely engage in waste handling and movement at the level of the household and therefore may not be in the best position to appreciate the effect of distance. Alternatively, perhaps it could be that they are physically stronger and do not mind carrying waste over longer distances. An internal analysis by sex of the respondents on the effect of distance on use of council provided waste disposal facility revealed findings as presented on Table 7.2.

Table 7.2: The effect of distance on the use of authorized waste collection sites

Sex	n / %	Strong Effect	Not Much	No Effect	Total
Male	n	41	14	7	62
	%	66.1	22.6	11.3	100
Female	n	100	23	1.0	124
	%	80.6	18.6	0.8	100
Total	n	141	37	8.0	186
	%	75.8	19.9	4.3%	100

$\chi2= 12.398$; df=4; P=0.015   *Household survey, 2013*

Table 7.2 shows men and women's responses to the extent of the effect that distance has on the use of council provided waste disposal facilities. About 66 percent of male respondents suggest that the distance in the location of the council authorized waste collection site

164

have a strong effect, an approximately 23 percent suggest that the effect is not much and 11 percent indicate it has no effect on their use of the waste facility. Among the female respondents, 81 percent say distance has a strong effect, 19 percent suggest not much and seven percent indicate that it has no effect. The difference in male and female responses are statistically significant ($\chi 2= 12.398$; df=4; P=0.015).

Cross tabulating the age of respondents with responses to the question about the influence of distance to determine use of public waste facility, analysis indicated that more than 50 percent of respondents of all age groups affirmed the influence of distance (Table 7.3).

Table 7.3: The effect of distance on the use of authorized waste collection sites: Distribution by age

Age	n / %	Yes	No	Total
16-20	n	11	10	21
	%	52.4	47.6	100.0
21-25	n	26	20	46
	%	56.5	43.5	100.0
26-30	n	32	29	61
	%	52.5	47.5	100.0
31-35	n	32	15	47
	%	68.1	31.9	100.0
36-40	n	19	16	35
	%	54.3	45.7	100.0
41-45	n	26	10	36
	%	72.2	27.8	100.0
46-50	n	11	8	19
	%	57.9	42.1	100.0
Above 50	n	20	7	27
	%	74.1	25.9	100.0
Total	n	177	115	292
	%	60.6	39.4	100.0

$\chi 2= 8.444$; df=7 P=0.295   *Household survey, 2013*

Table 7.3 shows that 68 percent and 72 percent of respondents within the age groups of 31-35 and 41-45 respectively indicate that distance has an effect on household waste disposal practices. As to the respondents who say that distance has no effect on their use of authorized public waste facility, the following age groups have noticeable results: 16-20 (48 percent); 20-30 (48 percent), 36-40 (46 percent) and Above 50 (26 percent). There is however no statistically significant difference in the affirmative or negative responses by age of respondents ($\chi2= 8.444$; df=7 P=0.295). This result implies that age makes no difference in the way respondents perceive the effect of distance on household waste disposal practices. Overall, these results indicate that there is a positive correlation between distance and the effective use of waste site, implying that, the closer the waste facility the more it will be used. Therefore, the location of council waste facility determines household waste disposal practices. This conclusion alludes to the expressed wishes of respondents. For example, when asked to make suggestions for improved waste management by the council to suit their requirements, 31 percent of respondents requested that more bins be made available. Requests were expressed in statements as:

"Every five houses should have a common bin accessible by the council vans for waste collection" [Female, post-secondary, aged between 26-30]

"Bins should be placed within reach and emptied regularly" [Female teacher, post-secondary, aged above 50]

"Council should increase the number of dust bins in the market" [Female tailor, secondary education aged 31-35]

7.3.2 Frequency of Municipal Waste Collection by the Council

Waste collection by the council is usually irregular and impacts on the condition of the waste site, the length of time households store their waste, and whether they look for alternatives to using the facility provided by the Council. Table 7.4 presents information on the frequency with which the Council collects and transports waste as viewed by household respondents.

Table 7.4: Frequency of waste collection by the Council

Frequency of waste collection	Count (n)	Percent (%)
Daily	19	7.3
Thrice a week	3	3.1
Twice a week	40	15.3
Once a week	146	55.7
Once in two weeks	1	0.4
Once a month	6	2.3
Never	42	16.0
Total	**262**	**100**
Missing	77	

Household survey, 2013

The results in Table 7.4 reveal that 90 percent of respondents say waste is not regularly collected, that is, at least once in two days as suggested by council regulation. Seven percent of respondents indicate that the public waste bin closest their homes is emptied by the council on a daily basis, 15 percent report twice a week and another 03 percent indicate thrice a week. The largest group, 56 percent, proposes once a week. These responses, especially of the last category of respondents confirm information from interviews with council authorities, which suggest that most of the city does not receive regular waste collection service. The result of this limitation is usually the disposal of waste at unauthorized sites such as along the streets, streams, gutters and overfull waste bins.

7.3.3 Condition of Waste Collection Site and Council Waste Management

In an attempt to evaluate the state of the waste collection sites, respondents were asked to describe the usual state of the waste collection site nearest to them. Results are presented in Table 7.5 and Plate 7.2.

Table 7.5: The state of the waste collection site/transfer station/public bin: Distribution by sub-division.

Sub-Division		State of waste collection site			
		Half-filled	Always filled	Always over-filled	Total
Bamenda I	n	2	2	0	4
	%	50.0	50.0	.0	100.0
Bamenda II	n	53	69	39	161
	%	33.0	42.9	24.2	100.0
Bamenda III	n	38	38	24	100
	%	38.0	38.0	24.0	100.0
Total	n	93	109	63	265
	%	35.2	41.1	23.8	100.0

Household survey, 2013

Plate 7.2: State of Council Waste Bin   *Source, Akum, 2015*

Table 7.5 shows that, 24 percent and 41 percent of respondents respectively, indicate that waste collection sites are always over-filled or always filled with waste. Spatial distribution by sub-division show that, waste collection sites in Bamenda I are always filled (50 percent) or half-filled (50 percent). In Bamenda II, 67 percent of respondents point to the fact that waste collection sites are usually overcome by waste. Forty three percent of this category suggests always filled and another 24 percent indicate always over-filled. In Bamenda III, respondents propose that 38 percent and 24 percent respectively that waste collection sites are "always filled" or "always over-filled". In general, only 35 percent of respondents suggest that the waste collection sites are "always half-filled" with waste. In line with results on Table 7.5 is the condition of waste collection site on Plate 7.2 indicating a full container of assorted waste. These results imply that, the waste collection sites are usually overwhelmed with waste pending collection. The overwhelming situation of waste at the waste collection sites in Bamenda II and Bamenda III can be attributed to high population densities as compared to Bamenda I with low population density.

In spite of this discouraging picture of the state of the waste collection sites and irregularity in waste collection, on average, the population believes that there has been an improvement in the state of waste management in the city when compared with the situation two years ago. Two years ago, when I met the Delegate to the Bamenda City Council, the Delegate had responded that it was also one of his major concerns and that he had promised to do something about it. In the household survey, respondents were asked to compare the current situation and that of two years ago. The results indicate that almost three quarters (73 percent) of respondents believe that there has been an improvement, 22 percent say the situation has not changed while 05 percent say the situation has deteriorated. An analysis by sex of respondents to situation indicates some differences between males and females in the percentage of change in all three-evaluation categories. This difference is however not statistically significant as indicated in Figure 7.2.

Household respondents have diverse indicators for perceived improvement in council waste management over the past two years. From the category of respondents who suggest that there has been improvement in management over the past two years, 49 percent of them argue that the city is cleaner than before; 44 percent announce that there has been improvement in waste collection and disposal services and the council; and 04 percent suppose that the council has provided communal waste bins thereby reducing the waste disposal in open spaces.

Figure 7.2: Evaluation of the state of solid waste management: distribution by sex

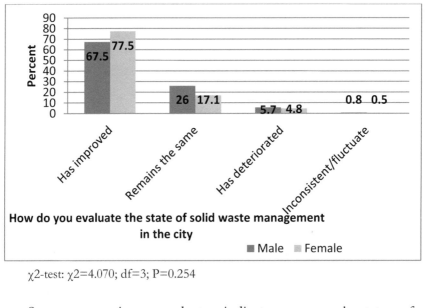

χ2-test: χ2=4.070; df=3; P=0.254

Some expressions used to indicate a general state of improvement from the code quotation grounding report in the content analysis are here presented:

"Heaps of dirt are not found on the street as before" [Female teacher, Post-secondary level, aged above 50]

"Rivers are free of refuse" [Male civil servant, secondary education, aged 16-20]

"Places do not smell as they used to be" [Female teacher,]

"The dump site at mile 6 Mankon has been cleaned" [Female, secondary, aged 31-35]

"Because standing water is no longer seen in some quarters?" [Female trader, post-secondary education, aged 21-25]

"The streets are being swept every night" [Male employed, secondary, aged 26-30]

"More bins available" [Male civil servant, post-secondary level, aged between 41-45]

"Waste now collected by refuse vans and disposed in areas far away from human habitation" [Female civil servant, post-secondary, aged above 50]

"At least they pass once a week, before they were never passing" [Female trader, secondary, age 26-30]

"Now the vans go into the quarter" [Male trader, post-secondary, aged 26-30]

The reasons advanced by respondents for general improvement in municipal waste management over the past two years range from increase in the staff strength and their efforts and improved service by the council. These opinions are presented in expressions as:

"Employment of cleaning staff" [Female, post-secondary, taxation officer, aged 36-40]

"Availability of dust bins" [Female, post-secondary education, teacher, aged between 31-35]

"Availability of council waste collection vans, waste deposit site and personnel to clear out waste into the van" [Male, office worker, post-secondary, aged 31-35]

"The council workers are doing their best to clean the town" [Female, retired teacher, post-secondary, aged above 50]

Another 9.8% attribute improvement to collaboration from the public in the use of waste facilities placed at their disposal, as well as improved awareness on environmental hygiene.

"The growing awareness on the need to have a clean environment" [Male, retired, above 50 years]

"The citizens are sensitized of pollution so they help the council to gather them" [Female retired worker, post-secondary education, aged between 41-45]

"By using the bins correctly when I am opportune to" [Female, aged 21-25]

"Collaboration of both council and quarter people" [Female administrator, aged 41-45]

Other reasons cited include elected officials efforts to fulfil promises made during political campaigns, clean-up campaigns, and recruitment of students during holidays to help with waste collection, penalties levied on violators and collaboration between the council and other waste operators. Some suggest that the council has come to understand the effects of poor waste disposal on the environment, such as floods resulting from clogged up drains, and public health, such as the prevalence of malaria and cholera, with increases in population in the city.

According to Figure 7.2, about five percent of respondents evaluate the waste management situation in the city as having deteriorated over the past two years. These respondents suggest that the situation has declined generally, due to inadequate service provision by the Council. These respondents suggested the following as indicators for deteriorated state of waste management: a general increase in the amount of waste generated; the environment is dirty; there is lack of collaboration from the public as people still dispose of waste haphazardly; the frequency of waste collection has declined; and clean-up campaigns are no longer used. Others complained of geographic and class discrimination, with some neighbourhoods favoured at the expense of others. This category of respondents makes comments such as the following:

"Increased waste quantities irrespective of increase in cleaning" [Female teacher, post-secondary, aged above 50]

"There is dirt everywhere" [Female housewife, postsecondary, aged 41-45]

"Overfull waste bin is not emptied in time" [male driver, post-secondary, aged 26-30]

"People still dispose dirt anyhow" [Female trader, post-secondary, aged 36-40]

"People are not willing to deposit waste in the bin" [Male farmer, post-secondary, aged 23-30]

"There are some quarters where they no longer go" [Female trader, post-secondary]

"Because most at time, the waste bins are often kept full for a long time" [Male farmer, post-secondary, 36-40]

"Absence of clean-up campaign [Female teacher, post-secondary, aged 36-40]

7.3.4 Public Participation in Council Waste Management Planning

The research results indicate that there is a low level of public participation in Council waste management planning. Decision making about waste collection strategies such as the types of bins to use, location of communal bins and the appropriate hours for the mobile van to pass to collect waste by the council does not take into consideration the opinion of stakeholders. The stakeholders of importance include the waste generators (households, market vendors and others), private waste operators and NGOs. To determine the level of public participation in Council waste management planning, respondents were asked to indicate who decides where the public bin, which they use, should be placed. To this question, 68 percent of respondents believe that the Council alone decides where the waste facility is situated. In addition, a remarkable 16 percent indicate that they do not know. The response of "I do not know" can suggest lack of interest in the affairs of the council by the population on issues that concern even their own wellbeing. It may also indicate that the Council ignores the population in decision-making. There is however, a possibility that most of the respondents really do not know who makes the decisions. They just assume that the council does and know that they, the respondents do not participate. It can be concluded that council waste management planning is far removed from the principles of good governance, which advocate transparency, accountability and a participatory approach.

7.3.5 Attitudes of the Waste Collection Team

The attitude of the waste collection team could affect the way households and other inhabitants use the waste facilities, particularly the mobile van pick-up provided for by the Council. Respondents were therefore posed the question, "How would you describe the attitude of the waste collection team/personnel to the public?" Results from the household survey are presented in Fig. 7.3.

Figure 7.3: Perceived attitude of waste collectors

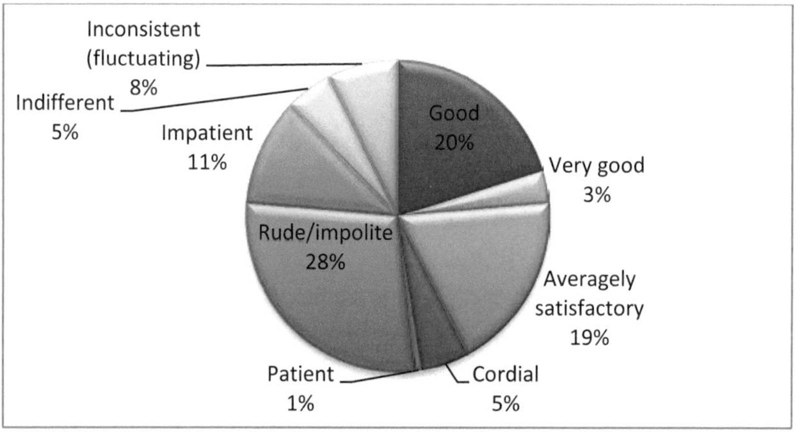

N=241

Figure 7.3 shows that 48 percent of respondents think positive of the waste collection staff attitudes. This category of respondents use attributes as good, very good, satisfactory, patient or cordial to describe staff attitude. On the other hand, 44 percent perceive their attitudes as negative, describing workers as rude/impolite, impatient or indifferent towards service users. Another eight percent consider staff attitude inconsistent. Further analysis of the results by the sex of respondents reveal no statistically significant difference ($\chi 2$-test: $\chi 2=8.367$; df=8; P=0.398). In any case, the fact that 44 percent of respondents see staff attitudes as negative has the potential to affect waste collection performance. For example, lack of patience for waste carriers to bring their waste to the mobile van discourages some households who tend to use alternatives as unauthorized disposal sites.

These results indicate a need for change in attitudes of waste collection staff that can encourage more households' waste transporters to use the Council mobile waste collection service. In this light, respondents were asked what suggestions they could make to waste collection team that could improve on household use of the mobile waste collection service. Some proposals are that the waste van should halt for much longer periods at each stop to accord people time enough to bring their waste. Other respondents add that the waste van should run very early in the morning and late in the evening. It should not be during standard working hours (8:00AM t0 3:00PM) to provide the opportunity for more people to use the service. During this period, household members working away from home must have left home.

7.3.6 Adaptability of Waste Equipment

The extent to which the structure (height, shape and size) of the waste bins provided by the Council and used by the public for MSW collection is adaptable to users can have effects on the success of the MSW strategy. In this light, household respondents were asked to appreciate the extent to which the waste bins are adapted to users in terms height, size and shape.

Table 7.6: Adaptability of structure of waste bin

STRUCTURE	Degree of Adaptability/Convenience							
	Adaptable		Uncertain		Not adaptable		Total	
	n	%	n	%	n	%	n	%
Height	134	53.6	36	14.4	80	32	250	100
Size	172	60.1	58	20.3	56	19.6	286	100
Shape	169	59.7	84	26.7	30	10.6	283	100

Household survey, 2013

Table 7.6 shows that more than half of the respondents perceive the structure of the waste bins as convenient on each dimension. Considering the responses for those who suggest that the structure of waste bins is not adaptable, 32 percent make a complaint against the height, 20 percent against size and 11 percent against the shape. In addition, a noticeable percentage indicates that they are uncertain. This can reflect an "I don't care attitude" from the population. It may also mean that these people do not use the waste bins. Such distant attitude or opinion to Council initiatives by the population can imply that the Council has very big latitude to handle the affairs of the public without fear or favour from the user of council services. Generally, the perceived problem is not the structure (size, shape or height) of the waste bin but the low frequency of emptying them. Caution should be given to these results in Table 7.7 because respondents were adults and their responses may not adequately reflect the thoughts of the main users who are children.

7.4 Assessment of the Bamenda City Council (BCC) Gender Policy

This section looks at the extent to which council resources and waste delivery policies and strategies are responsive to the gender practical needs and gender strategic needs of the households. The goal is to determine what gender gaps there are in the BCC waste management policy and strategies that can be addressed to enhance management, reduce women's workload and improve women's lives. To attain this goal, a gender policy assessment of the BCC is done using the Social Relations Approach by Kabeer (1994).

7.4.2 Gender Policy Assessment

In an attempt to determine the extent to which the waste management policy, strategies or practices are gender sensitive, we use the Social Relations Approach Concept, developed by Naila Kabeer (1994 cited in March *et al.* 1999) that focuses on institutional gender policies. Kabeer classifies institutional gender policies into two main types, depending on the degree to which they recognize and address gender issues. Institutional policies can therefore be gender blind or gender aware (sensitive). Gender blind policies disregard the distinction between sexes. As such, policies incorporate biases in favour of existing gender relations and may therefore tend to exclude women.

Institutional gender aware policy approaches recognize that women as well as men are development actors, and are constrained in different, often unequal ways, as potential participants and beneficiaries in the development process. Gender aware policy approaches on their part are further categorized into gender-neutral policies and gender sensitive policies (gender specific policies and gender redistributive policies). Gender-neutral policies recognize gender differences in a given society and seek to overcome gender biases in development interventions. Gender-neutral policies benefit both sexes in meeting their practical gender needs within the existing gender division of resources and responsibilities with no inkling of transformation. However, as cautioned by Moser (1993), gender neutral policies tend to gender blind development actions for which women's needs, especially Strategic Gender Needs (SGNs) are often ignored. Gender specific policies are conscious of gender differences, and interventions are intended to respond to the practical gender needs of either men or women within the existing gender division of resources and responsibilities. Gender redistributive policies on their part address interventions intended to transform existing distribution of resources and responsibilities to create a balanced gender relationship between men and women. They address both practical gender needs and strategic gender needs. As such, interventions create supportive conditions for women to empower themselves.

7.4.2 Gender Policy Assessment of the Bamenda City Council on Waste Management

In an attempt to assess the extent to which the BCC waste policies and strategies are gender sensitive, interviews were conducted with the BCC Delegate, the Chief of Service for Hygiene and Sanitation, mayors and councillors. Information from interviews reveals that the BCC has no official written document as a local policy for waste management. The BCC bases its work on the 1996 national Environmental Law and the MINATD 2004 Decentralization Laws 17, 18 and 19 that determine the functioning of councils, competences and resources. In the absence of a written policy, assessment relied on the strategies and practices actually used by the BCC to collect and dispose waste to carry out a gender policy assessment.

Table 7.7: Gender sensitivity analysis of BCC waste collection strategies

BCC Waste Collection Strategies/ Practices	BCC Gender Assessment		Policy/Strategy
	Gender Blind	Gender Neutral	Gender Sensitive
-Location of waste collection site		•	
-Frequency of collection by BCC		•	
-Regularity of waste collection service		•	
-Schedule of mobile waste collection hours	•		
-Adaptability of stationary bins to users	•		

Source: Household Survey, 2013

Based on a gender policy assessment of the BCC (Table 7.7), waste collection strategies are either gender blind or gender neutral. None of them is gender sensitive. The absence of gender sensitive policy implies no recognition of gender issues in municipal waste management strategies that can address and transform women's SGNs and subordinate position in society. Gender-neutral strategies on waste management are concerned with BCC interventions that target a clean city. These interventions include location of waste collection site, frequency of waste collection by BCC and regularity of waste collection service. This implies that council waste collection strategies to some extent meet the practical gender needs of women, who like children are those most involved in household waste management activities of storage and transportation to waste collection sites. The practical gender needs (PGN) are met by providing waste collection facilities like the stationary dustbins and the mobile waste collection service. By so doing waste accumulation in the domestic environment, which could be health threatening, can be evacuated. In addition, waste collection by the mobile waste team reduces the distance household members must cover to dispose of waste, allowing women and other household members involved in waste evacuation to devote more time to other activities. It is important to note that interventions, which meet practical gender, needs also have the potential to address strategic gender needs in the long run. This is because time and resources gained from spending less time on waste handling activities and poor health conditions of household members that could have been caused by the accumulation of waste can be used for gainful employment.

The strategies of mobile waste collection and the presence of stationary bins serve the practical gender needs of the households, but have been rated as gender blind because the time schedule for collecting waste is not sensitive to household and women's availability. As shown in later sections, respondents complain that the BCC waste collection schedule is not made public and not regular. The mobile team passes sometimes during working hours when most household members are out of home. This means that women unlike men who are likely not in productive activities in the public sphere will be responsible for transferring waste. In general, the mobile pick-up van technique seems to disfavour households were most adults, especially women work out of home during daytime. Respondents would prefer that the mobile waste pick up van passes very early in the morning or at late evening hours. Equally, the waste collection containers as the public dustbins are far above the height of some users, especially children. This handicap helps explains the reason for depositing waste around the trashcan and not into it even when it is not filled. In general, the BCC current waste management strategies do not meet strategic gender needs that call for empowerment and involvement in decision-making. This is explained by the fact that the council does not motivate proper waste handling efforts and does not involve households in their waste collection planning as presented and discussed in the sections that follow.

7.5 Environment and Gender Implications of Council Waste Disposal Techniques

Waste management practices can influence city hygiene, the environment and public health. The environment and public health impact of poor waste disposal have implications for women's gender roles. From this perspective, household respondents were requested to describe the way they perceive the general hygienic conditions of the city using attributes as very good, good, poor and very poor (Fig 7.4).

Figure 7.4: Perception of hygienic condition of city

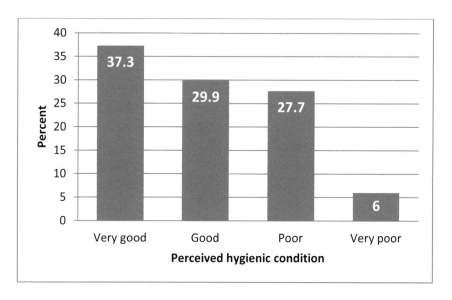

Fig. 7.4 shows that about two-thirds (66 percent) of the respondents have a generally positive view of the city hygienic conditions; 37 percent believe that the general hygienic conditions of the city are very good and 30 percent think the conditions are good. On the other hand, 34 percent of respondents perceive the condition to be wanting; 28 percent describe conditions as poor and six percent as very poor. When asked what indicator they used to determine their perception, most of the respondents who perceived the hygienic conditions of the city as positive cite the sweeping of streets. Such responses compliment council's efforts at engaging private contractors to do a daily cleaning of the town by sweeping major streets. Those who perceive hygiene negatively indicate that, though there has been an apparent improvement in waste collection efforts and general cleanliness of the town as compared to two years ago, it is limited to the main streets of the town. Comments of this nature

concur with the evaluation by the chief of service for hygiene and sanitation of the council and councillors who say that the interior parts of neighbourhoods hardly receive waste service from the council. She adds that the council has been contemplating of reinstating the monthly clean-up campaign strategy that had been halted after subcontracting cleaning to private operators.

7.5.1 Waste Disposal Practices and Effects on the Environment and Women's Roles

Waste disposal in water endangers the environment, threatens public health, increases women's workload and jeopardizes their income sources, as waste is often thrown into water that is used for domestic purposes and urban agriculture. When asked if household respondents have ever witnessed waste disposed in the rivers and other waterways, 69 percent answer in affirmation. Furthermore, information was solicited from respondents on whether the water in which they have witnessed waste dumped into is used downstream. To this question, 89 percent of respondents indicate that they have witnessed waste dumped in nearby streams used in the immediate area or downstream for drinking, washing, cooking and gardening. A spatial analysis by subdivision in the uses of susceptibly contaminated water downstream is presented in Table 7.8.

Table 7.8: Uses of potentially contaminated water downstream by sub-division

| Sub-Divisions | | Uses of water | | | | |
		Drinking	Washing	Cooking	Gardening	Total
Bamenda I	n	4	10	4	7	25
	%	16.0	40.0	16.0	28.0	100
Bamenda II	n	2	95	6	75	178
	%	1.1	53.4	3.4	42.1	100
Bamenda	n	3	50	6	34	93
III	%	3.2	53.8	6.5	36.6	100
Total	Count	9	155	16	116	296

$\chi2=35.500$; df=6; P<0.001   *source: Household survey, 2013*

Table 7.8 reveals that there is a statistically significant relationship ($\chi 2=35.500$; df=6; P<0.001) in the uses of potentially contaminated water in the different subdivisions. In general, the use of water from streams and rivers for drinking and cooking is low as compared to washing which is high. There is a difference in water use for gardening in the different sub-divisions. The use of water use for gardening is relatively higher in Bamenda II and Bamenda III than in Bamenda I. This can partly be explained by the presence of small scale farmers especially women, in Bamenda II and Bamenda III, who carry out market gardening activities, particularly in the dry season, when profit is highest. When water sources are polluted from waste, household members must travel long distances in search of water for domestic use or crop irrigation. The effect is increased work for those charged with such duties, in most cases women and children. In addition, water shortages caused by pollution can lead to low farm yields, discourage farmers and even destroy sources of livelihood. The economic activities of women are particularly impacted because they cannot use polluted water for family gardens and other women's traditional roles. Worse still some farmers whose interest for money is paramount will use the polluted water for cultivation, with implications for pubic heath. If using polluted water leads to illnesses like cholera, dysentery or food poisoning, women become further burdened with care giving roles.

Respondents were asked to indicate what sex (male or female) makes more use of the stream into which they had seen waste thrown. To this question, 84 percent of respondents propose that women use the water sources more than men do. The response simply confirmed traditional gender division and women's reproductive roles. This implies that any harm to water sources will seriously affect women adversely. This is one of the reasons why MSWM strategies and systems must take into considerations the interests of all actors, especially women, implicated in waste initiatives. Given that about three-quarters 74 percent of residences are closed to drainage sources (river, stream, swamps, gutters, etc.), there is a high potential for use of waterways as a major disposal site for waste in the city of Bamenda, with immense negative implications for the environment, public health and women's work load

7.5.2 Location of Communal Bins and Municipal Landfill

Inappropriate location of waste collection sites in relation to houses, catering and other business premises can have public health and environment implications. From this perspective, household respondents were requested to give their opinion on how appropriate they judge the locations of homes, cooked food/eating places, fresh meat sale stands and fruits vending places conditioned for immediate consumption (like peeled oranges, pawpaw and pineapples) in relation to the site of the public waste bin. Results are presented in Table 7.9

Table 7.9: Appropriateness of location of communal waste bin

Premises	Degree of appropriateness of location of waste bin				
	n/%	Appropriate	Inappropriate	Uncertain	Total
Homes	n	143	133	24	300
	%	47.7	44.3	8	100
Cooked food	n	65	178	28	271
	%	24	65.7	10.3	100
Fresh meat	n	64	164	39	267
	%	24	61.4	14.6	100
Instant edible	n	52	183	32	267
fruits	%	19.5	68.5	12	100
Total	n	**324**	**658**	**123**	**1105**
	%	**29.3**	**59.5**	**11.2**	**100**

Multiple Responses   *source: Household survey, 2013*

Table 7.9 generally suggests that the location of the premises identified is considered by more than half (60 percent) of respondents as inappropriate. Another 29 percent view the locations are appropriate while eleven percent indicate uncertainty. In terms of the location of homes from major waste sites, 48 percent hold that it is appropriate, 44 percent indicate inappropriate and eight percent are uncertain. The danger is that dust particles, and toxins from waste sites can contaminate foods and expose consumers to health risks when they are spread by wind or insects.

Interviews with appointed and elected the municipal officials, including the City Delegate for Bamenda, the Chief of Service in charge of hygiene and sanitation and BCC drivers, revealed that the city has one 'landfill'; however, a visit to the site does indicates that it would better be referred to as an open dump site. This dump site emits offensive odours and dust particles, which spread over long distances with the help of the gusty North East Trade Winds often referred to as the "harmattan". Fires characterized by thick smoke are frequent. They are purposefully set to reduce the volume of waste through burning of flammable materials in the waste stream such as plastics, tyres, batteries, dry leaves and grasses. The smell, smoke and dust cause choking and irritation to eyes, nose and throat and are likely to have negative health effects on the always-present scavengers, waste workers and households, which are not very far off.

The gusty prevailing winds contribute in spreading the smells and dust from the landfills (dumpsite) to settlements. Fires at dumpsites sometimes spread to neighbouring farms, bushes and soil, causing environmental degradation. Toxic dust particles settle on crops (vegetables and fruits) and contaminate them. If these foods are not properly washed before consumption, health risks are likely to result. Leachate infiltrates the soils and ground water, particularly the flood plains of the River Mezam and its waters, due to the absence of the liners normally used to prevent infiltration from sanitary landfills. At Mile 6 Mankon, for example, a family abandoned their house in 2011 because of the lamentable effects of the municipal dumpsite which was then located there. Uncompleted structures were abandoned and development around the neighbourhood was interrupted. In addition, waste spilled over into the main road, disturbing the flow of traffic. At the Bagmande dumpsite, population in adjacent places was already complaining of its inconveniences. For instance, Mr Angwafor Tse, who met this researcher in the field, said the stench and smoke from the dumpsite was actually a health nuisance. He added that most persons in the neighbourhood frequently suffer from cough, which was not the case before the dumpsite was located there. He advocates relocation of the dumpsite away from the present site.

The findings on households and Council waste disposal practices, the location of landfills (open-dumps) and the method of waste reduction are justifications and explanations of the Treadmill of Production Theory by Schnaiberg (see Chapter two – Section 2.7). Urban dwellers and the Council use swamps and other watercourses (particularly the Mezam River) to dispose waste. The Council has no sanitary landfill and practices open burning at the open-dump site as a waste reduction technique. Leachate from waste flows underground into water sources, which are consumed downstream for multiple domestic uses with implications for public health. Motor accidents have occurred as roads have been narrowed by over-filled dumps that extend to highway and visibility blurred by smoke as the case of Mile 6 Mankon. Surrounding vegetation and crops have occasional been burnt due to accidental fires from dumpsites. Pungent odour from decaying waste material and burning is believed by citizens to be the cause of common illnesses as malaria and respiratory infections. As Schnaiberg postulates, the government only intervenes when the impact of environmental destruction becomes alarming. In the case of BCC, the dumpsites have been relocated from Mile 6 Nghomgham Mankon to Bagmande in Mankon both in Bamenda II Subdivision, and currently at Mbelewa in Nkwen in Bamenda III Subdivision.

Summarily, this chapter has presented findings on the assessment of BCC waste management strategies. Waste management is one of the public utility services of the Bamenda City Council administration, which is carried out by a unit of the Department of Hygiene and Sanitation. This service has the responsibility of keeping the city clean by managing municipal waste. Two main strategies are employed: the mobile waste van and the placement of stationary communal bins. General hygiene and sanitation in the city has improved in recent years, but waste collection and disposal by council still lags far behind waste generation. Consequently, secondary waste collection sites are always overwhelmed by waste. The number of authorized communal waste sites/bins is inadequate, and they are distant from most homes. The result of this handicap is the high level of waste disposal in waterways and other illegitimate sites, with implications on the environment, public health and women's workload.

An institutional gender policy analysis shows that the BCC has no written waste management policy. The BCC however has waste management strategies and practices, which are assessed to be generally gender neutral. This implies that women practical gender needs are met to some extent with the provision of waste collection facilities by the council. Nevertheless, the absence of gender sensitive policies with interventions that target specific women and household waste needs mean that women strategic gender needs cannot be met through BCC waste management planning. As a follow up, women cannot be empowered because traditional gender division of resources and responsibilities are in no way addressed. It is believed that, motivating proper waste handling or penalising defaulters of waste disposal regulations will add value to waste management activities and benefit women who are the ones most often involved or affected by poor disposal effects.

Chapter Eight

Discussions, Conclusions and Recommendations

The main objective of this study was to examine municipal waste generation and management policies and strategies, and the gender issues and gaps implicated. To attain this goal, the strengths and weaknesses of the current waste management strategies were identified and examined to find a justification for mainstreaming gender in council waste management for better waste service delivery efficiency. In line with the research objectives and research questions, this chapter discusses the major research findings in relation with existing research literature and discourses on gender, environment and waste management. Conclusions are made on the extent to which municipal waste management strategies are effective, and their impact on the environment and women's roles. The chapter closes with proposals for measures to improve on municipal waste service delivery in terms efficiency, particularly through gender mainstreaming, and recommendations for further research. Material in this chapter is presented under discussions, conclusions and recommendations.

8.1 Discussions of Findings

8.1.1 Demographic Profile of Households

Objective one of this study sought to describe the demographic profile of households and household respondents in the Bamenda City Council (BCC). A majority of the household respondents are within the age range of 21-45 (77 percent). This is the active and potentially reproductive age group with household decision-making powers that can influence household waste generation and management. Considering income, a majority of households (63%) studied earns less than 100,000 FCFA a month. This is a relatively low average when compared to other city dwellers like Douala and Yaounde, which is averagely 200,000FCFA (Sotamenou, 2010). This relatively low-income level can be attributed to the absence of major companies and industries that are big employers. There was a statistically significant difference ($\chi 2=33.385$; df=8; P <0.001) in the monthly income levels of households in the three subdivisions of Bamenda I, Bamenda II and Bamenda III. Household incomes for Bamenda I inhabitants are considerably higher than for Bamenda II and III. Inhabitants of Bamenda I are mostly civil servants of a relatively higher socio-economic status. A positive correlation (.699) exists between the level of education of father and that of mother. This indicates that educated persons tend to marry one another. There is equally a positive correlation between household income and household size. This relationship implies that there is a tendency for people to depend on others. Such is popular in Cameroon and Africa where extended families are common. Again, another positive correlation (.392 and .390 respectively) is established between household income and level of education of parents.

Surprisingly, the findings reveal no relationship between household incomes and waste generation quantities in weight. This finding validates literature by Afon (2007) who found out in the study on household waste in Ogbomosso, Nigeria that households in the low-income zones tended to generate more waste than those in high-income residential zones. However, it is important to note that though lower income class zones may generate more waste than higher income ones, the latter tend to produce more waste of diverse composition. This revelation falls in line with UNEP (2009) literature on municipal waste generation and characterization.

The second part of objective one sought to situate households within the physical environment. This is because it is supposed that the features of the natural environment in which households are found can influence household waste management practices and vice versa. The main features taken into consideration are roads as well as the drainage features. More than two thirds of roads used by households surveyed for this study are unpaved. More than one-third of households surveyed are exposed to at least one drainage feature. Conversely, the Council waste pick-up van mostly plies the paved roads. The consequence is that, waste service delivery is limited to households located close to the main roads. The outcome is waste disposal in open space, gutters and water bodies as streams and swamps, and dumping along the streets. This finding identifies with those of Fombe (2006) and Balgah (2002) on a study of solid waste dumping and collection facilities in Douala, and Solid Waste Disposal in Bamenda Urban Area, Cameroon.

8.1.2 Female Domination of Household Waste Generation Activities

Gender analyses of household waste generation suggest that men, women and children participate in household activities that generate waste. However, the female (adult female and girl child) tends to participate the most. This finding attests to the feminisation of household reproductive works that have the potential to generate waste and consequently the feminisation of household waste generation. Findings further validate literature by Thu (2005), and Muller & Scheinberg (2003) on the central role of women in municipal waste generation in Vietnam. The gender roles are reflected in household waste activities as the female adult and girl child carry out food preparation and sweeping of the domestic environment that take place daily, while the adult male and boy child take the lead role in yard clearing that is carried out averagely once a month. The activity of compound clearing may likely generate less waste than the other two. This is because urban vegetation is limited due to the fact that houses in the city are built very close to one another for want of space.

8.1.3 Household Waste Comprises Biodegradable and Non-Biodegradable Materials

Findings further reveal that household waste comprises biodegradable and non-biodegradable materials with a higher proportion of the former. This result echoes reports of Ngnikam, 2000; Furedy, 2002; ENCAPAFRICA, 2004; Achankeng, 2005; Cofie *et al*, 2010, Behmanesh, 2010 and Sotamenou, 2010 on waste studies in Africa that suggest that biodegradable waste constitutes 70 to 90 percent of municipal waste streams in African cities, and between 30 and 70 percent in cities of Asia and Latin America (Medina, 1998; Beukering *et al.,* 1999). Tanawa *et al.* (2002), UNEP (2009) suggest that though the organic waste fraction is high, its proportion in the general waste stream is experiencing a decline in favour of the inorganic waste material. These authors attribute this shift to increases in incomes, economic production and adoption of western lifestyles.

As with the findings of Afon (2007), in a waste study of different ecological zones in Ogbomoso, Ibadan in Nigeria, income did not seem to have any greater influence on waste types generated between Foncha Street, a high income neighbourhood and Ntambag, a low income neighbourhood, or Ndamukong, a middle income neighbourhood. This could be explained by the presence of low cost products of non-biodegradable material originated from China and the emerging economies of South East Asia at prices affordable by a large sector of the population. The assertion that increases in income has a direct relationship with increases in waste quantities may hold true on a macro scale and not on a micro scale as for households and small neighbourhood communities. The average per capita per day waste production in weight in the BCC is 0.71Kg. This figure falls in line with the findings of UNEP (2004) and the Environmentally Sound Design Capacity-Building for Partners and Programs in Africa (ENCAPAFRICA). Findings of these organizations suggest that the average waste generation rates in African cities range between 0.50 and 0.87Kg per capita per day.

8.1.4 Plastics Constitute a High Proportion of Non-Biodegradable Municipal Waste

Survey results from questionnaires and household on-site waste measurements and observations indicate a variety of waste types. These include organic (biodegradable) waste from different activities and inorganic (non-biodegradable) waste materials, some recyclable and some non-recyclable. Plastics constitute a high proportion of non-biodegradable waste from households. This high proportion can be explained by the fact that plastic based materials are in themselves used as waste storage containers by over 70 percent of households and for packaging by over 90 percent of market vendors who parcel things for their clients. This finding on the dominance of plastic waste alludes to the potential environmental dangers that can be caused by this waste material.

The finding further provides a justification for the law by the government of Cameroon aimed at minimizing the effects of plastics on the environment. To this effect, the government of Cameroon through the Ministry of Environment and Nature Protection, and the Ministry of Commerce bans the importation, production and use of plastics as decreed by the joint MINEPDED/MINCOMMERCE, 2012 ministerial arête. This legislation, which was earmarked to take effect on 26 April 2014, caused some discontent among citizens, as alternatives for plastics were not yet within the reach of most Cameroonians. Currently, biodegradable plastics have been introduced in certain areas at relatively more expensive prices to the average Cameroonian. However, the law if fully implemented will reduce the plastic waste proportion in the waste stream and create a plastic free environment in the long run.

Most studies on municipal waste in the developing world and especially in Africa have paid much attention to the high fraction of organic waste in the waste stream and have often recommended the need for composting (Lardinois & Marchand 2000; Medina, 2000; Furedy, 2002; Mbuligwe *et al.*, 2002; Cofie, 2010; Sotamenou; 2010). The results of this study corroborate their findings and recommendations on the generation and management of organic waste. It however goes further to make visible the presence of plastic in the non-biodegradable content of the municipal waste stream, which poses impending environmental problem. Plastics remain a waste material, which can serve as a resource for recycling and by that offer job opportunities for waste pickers and scavengers in the informal sector in developing countries as advocated by Medina 2008.

8.1.5 Household solid waste separation is not common practice

Waste separation is less practiced. More than half the number of households within the city does not separate their waste (58.5%). This finding affirms those of Sotamenou (2010), Achankeng (2005) and Fombe 2005) who suggest that waste storage in most households do not reflect separation. All waste types are put in same containers. The absence of waste sorting at the level of the household further compounds the problems of waste treatment by waste management actors. As such, any waste management policies and strategies that fail to articulate the need for sorting is likely going to be inefficient especially in cost of composting and recycling efforts and the provision of municipal waste resources.

8.1.6 Primary Household Solid Waste Management is gendered

The activities of waste storage, separation and transportation at the level of the household are gendered. Waste is most often stored in plastic based material containers of varied sizes with or without lids. Waste separation as indicated in section 5.2.1 is not practiced by more than half (59 percent) of the number of households sampled. In households were waste separation is practiced (41 percent), the responsibility is highly feminized judging from the gender of the one who separates waste and from the degree of frequency to determine the level of involvement. The role of the female in waste storage and waste separation reflect their reproductive roles as articulated by Moser (1993) and Thu (2005).In addition, there is a close relationship between the waste generator and the sorter. This relation confirms the argument by Enger & Smith (1998) who suggest that the best manager of waste as regards storage and sorting at the source of production is the producer.

Children tend to be the ones most involved in transporting waste to the waste collection site. This finding echoes those of Fombe, (2005), and YRI & Pham Bang (2003) who studied substandard housing and slum development in Douala, and the role of children on the waste economy in Vietnam respectively. In Bamenda, the role of children beyond the secondary waste collection point is not visible. This is unlike the case in some countries of Latin America and Asia (Medina, 2008) where scavenging and waste picking that recruits persons of all ages is very prominent.

8.1.7 Spatial Discrimination in Council Waste Delivery Services

Beneficiaries of waste collection services by the council are limited to inhabitants who live close to main access roads in the city. This finding is similar to those of Fombe (2005) and Sotamenou

(2010), who focused on waste management in Douala and Yaounde respectively. Good roads and streets are indispensable for effective and efficient waste management (Achankeng, 2005). Bamenda, though endowed with an elaborate network of roads, most areas remain un-served by the council waste collection system. This is due to the poor nature of the roads, some of which are not passable particularly in the rainy season. Some inhabitants live in areas served only by footpaths, and some roads are very narrow due to the poor implementation of town planning regulations, as suggested by the Regional Delegate of Urban Development for the North West. Such conditions of roads make it difficult for council trucks to penetrate the interior of neighbourhoods. Consequently, the introduction of more communal bins by the council into the interior must be accompanied by accessible roads to bring any meaningful change in council waste evacuation efforts.

8.1.8 Contradictions in Environmental Consciousness and Household Waste Practices

Despite a very low level of waste sensitization by the Council and Delegation of the Environment, which is formally charged with this responsibility, the level of environmental consciousness (awareness) about waste in Bamenda is high. Inhabitants are generally aware that poor waste disposal has negative effects. This knowledge, however, does not translate to environmental sound practices in waste management. This disparity in knowledge and practice is exemplified in household waste disposal techniques and sites.

Open dumping and burning are common disposal techniques. More than half of the number of households studied dispose their waste at unauthorized and environmentally, and public health threatening sites as streams and gutters. Households dispose of their waste at diverse locations, with most of them using illegal sites, including streams, gutters, roadsides, open space, farms, etc. This result corroborates findings of earlier studies in other African cities as, Cotonou (Dedehouanou, 1998), Ibadan (Onibokun, 1999) and Yaounde (Sotamenou, 2010) that focused on municipal waste management issues. The dumping of waste at unauthorized sites has implications for management and the environment as discussed in the section that follows. This contradiction in the level of environmental consciousness and waste disposal practices resonates findings of other researchers on other environmental issues. For example, advocates of ecofeminism (Littig, 2001; Mies, 1993 Tiondi, 2001; Fonjong, 2008; Ngassa, 2013) argue that women have a closer relationship with the environment than men do. It is therefore expected that because women are highly dependent on this environment for their livelihoods as is the case in most rural Africa and Asia, they will tend it better. On the contrary, their farming techniques such as bush burning, slash bury and burn and cultivation along the slopes all go to destroy the environment on which they depend. This study on urban waste management reveals no significant difference neither in the women and men's perception of waste nor waste disposal practices.

Other studies on waste management in cities of Africa and the developing world have advanced findings on poor waste disposal policy and made recommendation for sensitization (Medina, 1998 & 2000; Buhner, 2012; Afroz *et al.* 2010; Konteh, 2009). This study articulates the fact that sensitization that ignores the gender factor in generation and management will remain at the level of raising waste consciousness without sound waste management practices. This is because there is a close relationship between household gender divieion of labour with impliations for municipal waste generation and management are gendered.

8.1.9 Market Waste Generation and Characterisation

Study results suggest that the tendency for a market to generate waste of different types and quantities is linked with the nature of products sold, the material used as packaging for items that enter or leave the market site, pastime activities carried out by vendors while at the market place and the size of the market. From this background, this study situates markets as major sources of commercial waste that adds to municipal waste as equally advanced by Heeramun (1995) and UNEP (2009).

There is a spatio-temporal distribution in the type and amount of waste that markets within the municipality generate. More waste is generated on Fridays and Saturdays during which days vendors furnish their shops with products, in preparation for major shopping often done by city inhabitants. Waste is generated as wholesale packaging is broken down for retailing activity. Furthermore, more organic waste is observed in the wet season, a period during which fresh fruits, vegetables and foodstuff like maize and beans ripen. This finding corroborates research by Achankeng (2005) and Afon (2007) that biodegradable waste content in the municipal waste stream increases in the months of April to October with the emergence of more fruits and vegetables. Though this study did not carry out density measures, it is likely that waste density amounts for Bamenda as Yaounde, Douala and Manila will reflect the findings of other tropical cities (Anschutz *et al.* 1995; Vermande *et al.*1994; Ngnikam *et al.*, 1997 and Tanawa *et al.*, 2002). Furthermore, this time and space distribution in the quantities and types of items sold in the market and waste generated there is reflected in household consumption and waste generation patterns.

Unlike households (see Section 8.2.8), a majority of market vendors (83 percent) dispose of their waste in the communal bins provided by the council. This can be attributed to the fact that bins are located on the market premises and therefore within walking distance. In addition, open spaces allowing scrutiny by neighbours and the fear of paying fines combine to discourage most persons from reckless waste dumping. This response by the vendors is a clear indication that municipal waste strategies influence waste disposal practices of users of waste delivery services.

8.1.10 Relationship between Council Waste Strategies and Household Waste Practices

Council waste management strategies influence households and market vendors waste handling practices. This result is consistent with the arguments by Sotamenou (2010), Achankeng (2005), Fombe (2005), Furedy, 2002 and Dedehouanou (1998), who attribute poor waste disposal practices by urban dwellers to inadequacies in council waste management strategies. Some of these inadequacies include the fact that the schedule for waste collection by the mobile team is irregular and not known by the population. The numbers of communal stationary bins provided by the council are insufficient.

The attitude of the waste collection team is considered by an influential proportion of respondents as unfriendly. The BCC lacks regulations (incentives and penalty), and education on waste generation and handling at the household level. These factors and many more tend to impact negatively on the effective use of the mobile van waste collection technique and the stationary bin method. Consequently, the waste collection sites are always overwhelmed with waste pending collection and waste is dumped in streets, streams/rivers, gutters and street sides.

The findings reveal a low level of public participation in council waste management planning. International development discourses (UN-HABITAT, 2008) suggests that community participation in decision-making is a very important instrument for good governance, which in itself is pertinent for sustainable waste management. According to the UN-Habitat (2012), good governance solicits participation, openness, effectiveness, responsibility and accountability in community projects and service provision. This study shows that the City Council generally fails to consult the inhabitants and is not answerable to the public for its waste management deficiencies. The absence of the participatory approach is further reflected in the fact that councils are not involved in waste management at the source through education, collaboration, motivation and penalties on good and bad practices. In addition, major waste management decisions and actions are centralized at the level of the City Council, ignoring the inputs from the communes and households. Such limitations can only promote stagnation in waste management delivery services.

8.1.11 Municipal Waste Management Practices and Environmental Impact

Findings of this study confirm research (Nkwemoh & Lambi, 2014; Buhner, 2012; NAPE, 2008; UNEP, 1996; Ogawa, 1996) which shows that municipal waste management practices impact on the environment and public health. For example, Nkwemoh & Lambi (2014) attribute water shortages in Yaounde partly to the pollution of water sources from urban waste. Expounding on poor waste management around the Lake Victoria in Uganda, NAPE (2008) discloses that households and industries discharge their effluents directly into the environment. Buhner (2012) like Medina (2003) who comment on waste management in developing countries propose alternative strategies that can be investigated and tried. They suggest that waste pickers and scavenging should be encouraged and promoted to enhance recycling efforts. Communities where landfills and open dumps are situated should be compensated with the desired infrastructure by the people as a way of mitigating contempt.

Findings of this study justifies the supposition of the Treadmill of Production Theory by Schnaiberg that environmental issues are only attended to when the level of destruction and impact becoming startling. In Bamenda, poor waste disposal exhibits itself in blocked gutters, roadside litter and contaminated water sources, especially in the plains of Bamenda II. The Council's main waste disposal site is an open dump loosely referred to as the landfill. Open burning is the main technique of treating or reducing the amount of waste at the dumpsite. Smoke and pungent smells emanate from the public open waste dumpsite due to burning. This method of waste elimination is the least preferred method as classified by the Waste Management Hierarchy concept, which rates MSW elimination methods from the most preferred to the least preferred options: Prevention – Reuse – Recycle – Incinerate (with energy recovery) – Landfill – Dump – Open Dump.

Ongoing debates as cited in Buhner, 2012 and Best Practices on Solid Waste Management in Nepalese Cities (2008) presented in the NGO Practical Action Nepal attribute waste management problems and their environmental effects to multiple factors. These include rapidly growing urban communities, lack of cooperation from citizens on waste minimization efforts and payment for waste collection, staff shortages, weak public institutions and lack of cooperation between public and private sectors, and inappropriate and inadequate assistance from donor countries. For example, external support agencies limit resources on mandates, modes and scope under which they can operate projects. They sometimes ignore the resultant effects of projects that may require the management of waste. Buhner (2012) argues that international donors sometimes lack a full understanding of the socio-economic, cultural and political factors that influence the selection of a given waste management system. The result is that some donor agencies transfer home-adapted technologies to recipient countries, which are not equipped for them. These imported technologies in the end tend to be liabilities rather than assets to the communities and management. This challenge is well exemplified in Bamenda by the presence of dilapidated and abandoned automated waste equipment from Switzerland. There is constant breakdown given the poor state of roads and advanced age of vehicles. The problem is further compounded by the fact that human expertise as spare parts is not readily available. The equipment as indicated by the chief of service of hygiene and sanitation and some councillors is currently a liability and not an asset to the Council. The abandoned equipment suffers from the aggressive effects of the tropical weather conditions. While the long run environmental effects of this degraded equipment cannot be determined by this research, it has become an eyesore that spoils the aesthetics of the urban environment. The challenge is therefore to develop and promote disposal systems that require a minimum level of mechanical equipment as suggested by Buhner, 2012.

Findings of this study suggest a high proportion of organic material in household waste. This result reflects findings by Achankeng (2005), NGO Practical Action Nepal (2008); Medina (1998), Afon (2007), UNEP (2009) and other researchers on waste management in the South that about 65 to 75 percent of waste from urban areas is organic. Going by the dominant organic waste component of municipal waste, low level of industrialisation and prevailing urban agricultural activities, composting at large scale would be expected as a strategy by waste managers to handle urban waste. This, interestingly, is not the situation. According to Mensah and Larbi cited in Buhner (2012), all public composting efforts in Ghana have failed. This is also true of Ibadan in Nigeria (Onibokun, 1999). This study dares to commend that (in spite of failures in composting efforts in some cities), any form of composting that sets out to enrich the soil irrespective of the provider is much needed in Bamenda where urban agriculture is threatened by environmental degradation. Institutional capacity, political will and collaboration from all stakeholders is required.

8.1.12 Waste Management and Health Impact

Though this study could not scientifically determine the effect of waste disposal on health, it is established that the people think there is a relationship. This is concluded from discussions with inhabitants who live near open dumps. This finding is in line with literature (Dat, 1995; UNEP, 1996, Sriniva, 2012) that argues that improper waste handling at the primary or secondary level of waste collection and disposal can cause serious impacts on health, just as it does on the environment. Srinivas (2012) and UNEP (1996) reporting on health impacts of solid waste posits that poorly managed household waste can lead to the spread of infectious diseases as it attracts flies, rodents and other creatures that that can serve as disease vectors. Current waste discourses stress on the role of waste plastic as a health and environmental hazard. Literature reveals that in most industrialized countries, and increasingly in some developing countries, coloured plastics have been legally banned. In India, the Government of Himachal Pradesh has banned the use of plastics and so has Ladakh district. This call resonates the ban on plastics in Kigali in Rwanda and is presently the case in Cameroon as per the joint arête of the Ministry of Environment and Nature Protection and the Ministry of Commerce banning the importation, production and use of plastics (MINEPDED/MINCOMMERCE, 2012).

8.1.13 Municipal Waste Management Practices and Impact on Women's Gender Roles

An assessment of the BCC an institutional gender analysis of its waste management policy reveal that the Council has no written regulations that guide urban dwellers or even council's performance on waste practices. It further reveals that the BCC waste management strategies are gender neutral and likely to generate gender blind actions that often do not favour women (Kabeer, 1994). The results discussed above show that municipal waste management practices affect the environment and public health. The effects of the waste management practices have implications for women's gender roles.

This assertion is congruent with research by Thu (2005), Muller & Shienberg (2003), Chi (2003), Dat (1995) and Mies (1993) that points out the role of women in municipal waste generation and management and women's close connection to the environment and vulnerability to the impacts of poor waste disposal. Current gender discourses recognize the reproductive role of women in waste generation and management. This role is manifested in household reproductive works of food preparation, sweeping of the domestic environment, waste sorting, and transportation of waste to collection sites. As caregivers, women endure the most of taking care of children and other family members who suffer from diseases resulting from poor waste management.

Women's participation in the waste economy can also be a source of livelihood, but it is constrained by several factors (Dat, 1995; Chi, 2003; Thu, 2005). Women often participate in low paid occupations like sweeping public places due to low educational status when compared with men. It can be seen that, women turn to replicate their household reproductive roles in the public sphere. This is exemplified in the type of waste productive works (paid work) they do. Research findings in conformity with literature (Dat, 1995; Chi, 2003; Thu, 2005) show that women who work with contractors in collaboration with the council are mostly involved in sweeping while men are involved in automobile and high technology related movement of waste works. They rarely participate in scavenging at the landfills because they are often far away and insecure. Financial insecurity deprives them of equal access to and control of equipment, such as bicycles or motor bikes that facilitate movement and allow successful participation in the waste business. Unlike in South East Asia and Latin America where scavenging is done in large scale, this activity is yet to gain importance and recognition in Bamenda. The participation of women in the waste economy in Bamenda is limited to women passing by homes or market places to ask for or buy waste products like bottles for reuse or fruit peelings and left over food as animal food.

8.2 Conclusion

This study sought to identify gender issues in current municipal waste generation and management practices, and to make proposals for gender mainstreaming in the collection and disposal policies and strategies of councils. Specifically, the roles of women, men and children in household waste generation were examined. The study concludes that household waste generation is gendered and feminized. The results show household waste generation and management occur primarily within the sphere of women's reproductive role in food preparation and other domestic works. From multiple response answers, the adult female always participates (80 percent) in food preparation and 27 percent in sweeping of domestic environment. The girl child always participates (25 percent) in food preparation and 45 percent in sweeping of the domestic environment. This level of participation is contrary to that by the adult male and the boy child who are less involved. These are activities that are carried out with the highest frequency as oppose to yard clearing which is male inclined but done only monthly. More so, in urban areas where houses are closely built, there are hardly large open spaces requiring clearing. Other important findings relate household waste amounts and household waste types. Biodegradable waste material dominates the waste stream, while plastic is the most important component of the non-biodegradable segment of the waste stream.

Examining the contribution of different household members to waste handling activities, findings demonstrate that primary household waste management activities of sorting, transportation and disposal are gendered. Women are very involved in waste sorting and disposal, followed by the adult male, the girl child and then the boy child. The children (60 percent) particularly the boy child and adult female (30 percent) play the lead role in waste transportation from home to secondary waste collection site. The role of the adult male - particularly those who are heads of households - is very insignificant. Waste sorting is done in less than 50 percent of households surveyed. One of the reasons advance for not sorting waste is the absence of waste recycling facilities by the authorities that

201

be. The creation of recycling plants will exploit waste resources and create employment opportunities for a community that has a high level of youth unemployment.

The study also looked at the level of environmental consciousness and use of sound practices in household waste handling. The results reveal that majority of respondents are aware of the harmful effects of poor waste disposal to the environment and public health, but their waste disposal practices are nevertheless environmentally unsound. One of the key results realized is that a majority of households dispose their waste at illegal sites particularly road sides, streams and gutters destroying urban aesthetics and the environment. The study concludes that the distance from households to authorized secondary waste collection site influences household waste disposal practices.

For market waste generation activities and characteristics, the research results reveal that the quantity and type of waste generated is strongly related to the types of products sold and the initial packaging of retailed products. For instance organic waste materials as grass and plantain leaves come from fresh tomatoes packaging which is unpacked during the retail process. Waste quantities and qualities also vary among markets, the bigger the size of a market in terms of amounts and types of goods sold, the more waste it generates in volume and types. In addition, the pastime activities of vendors contribute to waste quantities and composition, and these are influenced by gender. For example, women vendors select vegetables, crack egusi (pumpkin seeds), peel potatoes and carrots in the market for sale or to take home for food. They do this to maximize time use because of the many roles they have as businesspersons and home managers. As is the case with households, a majority of market vendors (82 percent) do not separate their waste. However, unlike households, a majority of vendors (83 percent) do throw their waste into the communal waste bins provided by the council. This result shows that council waste collection strategies strongly influence the waste practices of urban dwellers.

The study also sought to determine the extent to which council waste collection strategies influence household waste handling and disposal practices. The results show that, although there has been some improvement in waste collection by the council within the past two years, waste collection strategies largely lag behind households' disposal needs. The distance of the authorized waste collection point, in particular, influences household waste disposal practices. Waste generation and handling practices at source are ignored. Households are comfortable with transporting waste at a distance of about 100m. However, only 41 percent of households are located within 100m of public bins. The numbers of bins are considered by the inhabitants as grossly insufficient. The actual waste collection schedule by the council is very irregular, especially on streets that are not tarred, and even worse in the interior parts of neighbourhoods. Public participation in planning municipal waste strategies is minimal. Households' opinions are not sought when planning. Consequently, many households dispose of their waste at unauthorized sites as streets, gutters and streams. The waste collection sites, particularly Council provided stationary bins are usually full or overfull. Littering is also very common. These poor disposal techniques tend to spoil urban aesthetics and the environment, and even threaten public health.

An appraisal of the effects of current municipal solid waste disposal strategies on the environment was done with implications for women's roles and livelihood. Results reveal a gender-neutral waste strategy with negative effects on the environment that have implications for women's roles. The shortcomings of the existing waste collection system have effects on women's household reproductive workload as they sometimes have to travel long distances to dispose of their waste. As a result, waste is sometimes stored at home for longer periods than it would otherwise be, causing odour and making the home environment often managed and used by the woman unpleasant. Toxic material in waste disposed off in water contaminates and make water unfit for drinking, laundry or irrigation. Plastics thrown in waterways take up the banks of streams and surrounding swamps rendering the farm areas not cultivable. By reducing potential of these farmlands, urban farmers (who usually are women) have their livelihood sources threatened and render them poor.

8.3 Recommendations

8.3.1 Recommendations to the Bamenda City Council

The Council should develop a more inclusive waste management system that involves the sources of MSW generation. Council waste policy and strategies which have been judged as gender neutral in this study, so far focused on waste collection, transportation and, to a limited extent, disposal, largely ignoring waste generation. An effective municipal waste system that aims for a sustainable clean environment and good public health must start with activities at the source, including the actors involved in waste generation and their interests. Rules or guidelines on types of waste generated and how they should be handled are important to stakeholders of waste management. This is because each actor and component of the waste management system has a role to play in its proper functioning, and because household gender division of labour affects MSW generation and management. Guidelines and regulations at the level of the household and markets should focus on waste avoidance and waste minimisation, waste separation and recycling, including composting, and reuse of waste. These waste management options reflect the preferred techniques identified by the Waste Management Hierarchy Concept. So far, Bamenda, like most cities in Cameroon and the developing countries, has regularly practiced open dumping and burning which are the least preferred options by the aforementioned concept. Suggested policy review will prevent the production of unwanted waste as plastics whose production, importation and use are currently banned by the government of Cameroon. Waste minimization will reduce the amount of waste due collection and cost of transportation to the dumpsite. It will also increase the life span of landfill/dump sites. Proposed policy guidelines that address waste types and proper handling techniques will reduce the effects of toxic or unhealthy waste on the environment.

A more organised waste economy promoted by the Council is needed. The waste economy is concerned with techniques, activities and actors that seek to transform waste from a liability to an asset. In most of Africa, Cameroon and Bamenda, large scale waste recycling, composting, scavenging and treatment techniques and activities are absent in council policy. These techniques and activities of management, which add value to waste and make it resourceful, provide employment and promote the local economy. The absence of legislation and regulation that promotes them implies that no funding, or very little if any, is solicited for them by the Council or other waste managers.

Environmentally sustainable practices by the Council, department of environment and other waste handlers must be encouraged. Legislation and practices governing the location, construction and use of landfills or open dumps needs revisiting. The participation of stakeholders of importance (households, NGOs, small-scale waste agents, etc.) and influence (Council and department of environment), and a sound vision of environmental ethics will be necessary in formulating such legislation. Such a policy will address the present weak position of public participation in the present waste management planning system and ensure good governance principles of accountability and transparency, as well as environmental ethics. The goal will be to create sanitary landfills that safeguard public health and protect the environment.

Gender mainstreaming in MSWM policies and strategies by the Council is required. Research results suggest that household waste generation and management are gendered, but municipal waste management is not gendered. The key role of women in primary level waste generation and management activities requires gender sensitive waste policies and strategies, and a gender sensitive council budget for waste management and disposal that will meet their practical and strategic gender needs. The Council should make the MSWM system responsive to women's needs. Research results show that the current number of communal public bins provided by the Council is grossly insufficient. In addition, the location of bins in terms of distance from households and spatial distribution is too far and not satisfactory. To these concerns, the provision of more communal bins spatially located within convenient use by the Council decided in collaboration with households is recommended. Regular collection of waste by the itinerant truck at known and respected schedules will also help to address women's practical needs by reducing the stress of long walks to far off waste collection sites. This is because women are generally known as domestic waste managers by virtue of their household reproductive role.

The Council in collaboration with NGOs involved in waste management issues should make waste and waste activities resourceful. Adding value to waste at source of waste generation can contribute to addressing Women's Strategic Gender Needs (SGN). This can be done by making policy that will add monetary value to household waste through incentives for waste separation. Such a policy will be appreciated and implemented by households. This is because research findings revealed that many households are ready to separate their waste if that action is remunerated. Furthermore, policy that will discourage waste disposal, particularly unhealthy waste into watercourses and swamps is needed. Such a policy if well implemented will ensure a clean environment and provide livelihood sources for women who are urban farmers. This is because women as main generators of household waste and domestic managers have access and control over most of household waste and handling activities. As such, women's strategic needs will partly be addressed and poverty reduced. This is in line with the gender analytical framework by Moser (1993) and Harvard Analysis Framework that seek to make visible the strengths of women and the discrimination against them through a gender disaggregated data analysis. This is an analysis that seeks to present the needs and benefits, interests and contributions of any given project by gender.

Providing waste dumps of reachable height for children by the Council is advised. Children play an important role in household waste transportation and disposal. Research results showed that waste is often dumped around the public waste bin even when it is not full. One reason for this may be the inability of children to get waste into high-walled bins. Strategy that takes into consideration that children play a principal role in waste disposal is very much needed. The conception, production and provision of communal waste bins should pay attention to the shape and height of the waste equipment and fashioned within reach of children.

8.3.2 Strategies to Improve on Municipal Waste Management

Study results reveal a gross inadequacy of Council's strategies to collect waste from the city in spite of increased efforts. An increase in the number of waste bins made available to urban dwellers and a distribution that ensures a good geographical coverage and takes into consideration the interior parts of neighbourhoods are recommended. Research findings show that Council waste management strategies and practices ignore the waste activities at the source of generation. By overlooking these activities, many households tend not to practice waste minimization techniques that can reduce the volume of waste. On the contrary, households produce unhealthy waste, which they store and dispose without sorting for collection and transportation to final dumpsite by the Council. As such, much of the Council budget is spent on waste collection, and yet most of the City does not feel the impact of its services. To encourage waste minimization and cost of waste collection and transportation by the Council, spontaneous as well as planned home visits be made by a team of council hygiene. The objective is to educate households on waste minimization techniques, encourage and emphasize on proper use of bins with lids and waste sorting. Incentives should be given to households who comply with rules in monetary value or equipment, while penalties are meted out to defaulters. Penalties could include payment of fines in monetary terms or cleaning up of any mess made.

According to research results, some household respondents suggest that their households do not separate waste because the Council waste collection sites have no provision for it. Sorted waste makes treatment easier, and encourages composting and recycling activities. These are waste management techniques and activities that reduce the amount of waste transported to the landfill, extend the life span of landfills, and provide livelihood sources. With a vision for recycling and composting at macro scale, councils should therefore provide facilities for waste separation at the waste collection sites to encourage users. The results also reveal the willingness of households to pay for waste collection. Council or other waste collection agents should capitalize on this to make waste collection a self-supporting and sustainable activity. By so doing, the cost of waste collection, which is currently borne solely by the Council, can be greatly reduced. Furthermore, any disposable funds provided by the payment initiative can be used by the Council to provide more waste facilities and services and made accessible to a greater population.

Findings suggest the absence of globally much applauded approach to development initiatives. A participatory approach to any community initiative (such as waste management) is highly recommended for success. Greater collaboration between the council and other stakeholders charged with urban environmental protection, such as the regional delegations of health, environments, housing and urban development, agriculture, education, and external funding bodies and should be undertaken. Research results indicate that over two-thirds of the respondents have hardly received any form of education from legitimated authorities as the Council and the department of environment on waste handling issues. This could be one of the many reasons for which households attach not so much importance on the need for proper waste disposal. Waste sensitization and environmental awareness education by the council and related departments are strongly recommended. These should be provided through every available media - print, television, radio, billboards, etc. In an attempt to reach out to many and internalise in the population a culture of a clean environment, and taking into consideration the principal role of children in waste disposal from homes, waste management issues should be included in the curricula of schools.

Due to female domination of household and market waste generation and primary level waste management activities, the role of women in ensuring a clean urban environment and public health should be accentuated to households, women, and children in particular. As mothers and educators at home and beyond, the participation of women in urban waste decisions and education about waste management is very important. Therefore waste management sensitization efforts must target women in places where they can best be reached – churches, women gatherings, markets, hospitals, and events as the Women's Day Celebration, etc. In addition, neighbourhood meetings should be held to discuss waste management issues. Such meetings must ensure that women have the space and the voice to articulate issues of concern. In addition the venue of such meetings must be accessible, and the time convenient (away from darkness) to ensure women's security. The members in each quarter should discuss and find alternative ways (other than council bins) of disposing their waste that are environmentally friendly, less costly and economically profitable for citizens. For example, people within the quarter could be identified who can make use of various types of waste materials, which will then be collected and made available to them. This would reduce the amount of waste that is destined for council collection.

Summarily, the quote from Muller and Schienberg (2003) aptly describes the contribution of women to waste generation and management and the impact of inadequate waste management services on women's reproductive work in most cities of the developing world and Bamenda in particular.

"Irrespective of the status of women outside of the household, within the home women are widely accepted as the caregivers, food preparers, and maintainers of the domestic environment. In most societies, this role carries over to an accepted role for women in community maintenance, often focusing there as well on cleanliness, health, and order. Therefore, any attempt to improve community urban services must logically take special care to consult women, who are almost certainly the ones most affected by changes or improvements. Taking household garbage to street corner dustbins may be easy, but it is not so easy when the distance between house and dustbin is too large. It is natural that children fall ill, the burden of caring for sick children who have been exposed to human faecal matter or vermin and disease in uncollected garbage falls disproportionately on the mothers, sisters, and grandmothers of those children"

In a majority of cities in the developing world, cultural norms and traditions compel women to domestic roles. Implementing gender mainstreaming in municipal waste management policies and strategies will go a long way to reduce the workload on women. In addition, making waste policies and strategies that can render monetary value to household waste will avail women even at the level of the household to gender reproductive waste works that will meet both their Practical and Strategic Gender Needs. The ultimate outcome of engendering waste policies and strategies will be improvement in waste collection and disposal that will make for a more sustainable clean city. Gender considerations in waste generation and management should therefore constitute an important component in managing the daunting solid waste situation of African cities experiencing waste problems similar to Bamenda and with limited resources. This research contends therefore that waste governance policies and strategies that take into consideration all sources of urban waste generation, all actors involved in waste generation and management and all interests implicated is much desired. Gender aware municipal solid waste management policies and strategies with increased human and financial resources will go a long way to address waste and other urban problems in African cities.

8.3.3 Recommendations for Further Study

Gender issues in waste management go beyond the household and market places and are implicated in the inadequacies of poor waste disposal, or the strengths of proper and lucrative strategies of management. This is because MSW collection and disposal techniques and strategies affect women's gender roles and gender needs. This study has focused on gender issues and gaps in household and market waste generation and primary management activities, giving only passing attention to gender issues in the waste economy, a sector that is not yet well developed in Bamenda. Future research should focus on gender inequalities in this economic sector in terms of income, cultural barriers, employment policies and health among others. Further research in these areas will provide more information on the role of men, women and children in activities of the waste economy (recycling and scavenging).

This research limited itself to waste generation in the household and markets. Other sources of waste, such as institutions, factories, auto repair premises, slaughterhouses and hospitals (that could generate more toxic or unhealthy waste) were not studied. Studies of waste from such sources could bring further information that will complement the findings of this study and previous researches, providing a well-informed basis on which a much more holistic municipal waste management planning system can be grounded.

The study was based on the Bamenda City Council and recommendations could be limited to this locality. Similar studies should be carried out in different cities such that results can be compared. With more findings, recommendations made can be valid nationwide and even beyond.

Bibliography

Achankeng, E. (2005). *Sustainability in Municipal Solid Waste Management in Bamenda and Yaounde, Cameroon.* PhD Thesis. School of Social Sciences, Department of Geographical and Environmental Studies (GES). University of Adelaide

Adams, K., P. S. Phillip and J. R. Morris (2000). "A radical new development for sustainable waste management in the UK: The introduction of local authority Best Value legislation." *Resources, Conservation & Recycling.* 30(3); 221-44.

Afon, A. (2007). "An analysis of waste generation in a traditional African city: the example of Ogbomoso, Nigeria." *Environment and Urbanisation, http://eau. sagepub.com* , 526-537.

Afroz, R., Hanaki, K. Tuddin, R. (2010). The role of socio-economic factors on household waste generation: A study in a waste management program in Dhaka city, Bangladesh. Volume 5, Issue 3, PP 183-190. Accessed: 21/07/2014

Akum. H. K. (2006). Diffusion of Innovations for Restructuring Small-Scale Farming: Case of Micro-Credit Financing in Mezam and Ngoketunjia Divisions. MSc Dissertation submitted to the Department of Geography, University of Buea, Cameroon.

Arrete Conjointe No. 004 MINEPDED/MINCOMMERCE du 24 Oct. 2012, Le Ministre De L'Environnement, de la Protection de la Nature et du Développement Durable.

Attahi, K. (1999). « Abidjan, Cote d'Ivoire. » In Adepoju and Onibokun, A.G., ed., *Managing the Monster: Urban Waste and Governance in Africa.* Ottawa. International Development Research Centre.

Ayuk, E. P. (2012, January 06). "Incentives should be provided". Cameroon Tribune, pp.11.

Balgah, S.N. (2002). Solid Waste Disposal in Bamenda Urban Area, Cameroon. Regional Conference of the International Geographical Union (IGU) 4-7. August 2002. Conference Proceedings

Batley, R. (1994). "The consolidation of adjustment: implications for public administration in public." *public administration and development* , volume 14, 489-505.

Beede, D. N. and D. E. Bloom (1995). Economics of the generation and management of municipal solid waste. Cambridge, National Bureau of Economics Research (NBER): 97.

Beukering, P. V., M. Sehker, R. Gerlagh and V . Kumar (1999). "Analysing Urban Solid Waste in Developing Countries: a perspective on Bangalore, India." *Working Paper* No. 24

BILL N0 762/PJL/AN on the Orientation of Decentralization No 51/AN7TH Legislative Period. Legislative Year 2004 2nd ordinary session (JUNE 2004).

Botkin, B & Keller, E (2000). Environmental Science: Earth as a Living Planet. New York: John Wiley & Sons, Inc.

Braathen, N. A. (2004). Addressing the economics of waste: An Introduction. *Addressing the economics of waste.* OECD. Paris: OECD: 7-22.

Brand, K. (1997). "Environmental Consciousness and Behaviour: The Greening of Lifestyles." In Redclift and Woodgate (eds). *The International handbook of Environmental Sociology,* Edward Elgar, Cheltenham, UK and Northampton, MA, USA, pp. 204-217.

Brown, T.E. (1991). "Waste Management Systems: An over view." *Long term waste planning and management.* T.E. Whitehead. Newcastle, University of Newcastle; Board of Environmental Science, *Occasional Paper.* No.17. 17: 27-62.

Bromley, Victoria, Baseline Analysis of Gender Sensitivity: IDRC Funded Projects 1995-1996, September 1996, IDRC/GSD Internal Document

BUCREP (2010).Bureau Central des Recensements et des Etudes de la Population (BUCREP). 3eme RGPH. Institut National de la Statistique du Cameroun. Yaoundé

Buhner, M. (2012).http:www.knowthe flow.com/2o13/3-alternative-ideas-for-waste management-in-developing-countries/

Care. (2004). Reclaiming Rights and Resources: Women, Poverty and Environment. Nairobi: Care International.

Coady International Institute (n.d.). A Handbook for Social/Gender Analysis. Canada. Covenant? Adelaid, EPA South Australia: 4.

CDC (2001).*Word Processing Data base and Statistics Program for Public Health*. Centre for Disease Control and Prevention. Epi Info 6 Version 6.04d.

Chi, Dang Kim (2003). "Women's Role in Waste Management in Vietnam: Gender and the Waste Economy-Vietnamese and International Experiences", M. Virginia and Nguyen ThiAnh Thu (co-editors). *National Publishing House, Hanoi,* (pp.69-74).Vietnamese Version.

Chuba, M. N. (1978). Refuse Disposal – Bamenda Town. Bamenda, Hygiene and Sanitation: 4.

Daly H. E., & Farley, J. (2003). *Ecological economics. Principles and applications*. Washington, DC: Island Press,

Dankel. I. & Davidson. J. (1989). *Women and Environment in the Third World: Alliance for the Future*. London: Earthscan Publications Limited.

Dat N. X. (1995). "The Health Impacts of Waste Collection on Women in Hanoi." In D. Baxter (ed.), *Gender, Environment and Development in Vietnam*, National Political Publishing House, Hanoi, (p.249).

Decree No 2005/088 of 29th March 2005 organizing the Ministry of Women's Empowerment and the Family.

De Tilly, S. (2004). *Waste generation and related policies*: Broad trends over the last ten years. Addressing the economics of waste. OECD. Paris, OECD. 1: 23-28.

Dedehouanou, I. H. (1998). "Coping with house waste management in Cotonou." *Environment and Urbanisation* , Vol. 10, No.2. 191-2008.

DEFRA (2011). Department for Environment, Food and Rural Affairs Nobel House, 17 Smith Square, London, SW1P 3JR. Government Statistic Service

Diana Lee-Smith, Nancy karanja, Mary Njenga, Thomas Dongmo, Gordon Prain. (2010, april). "The role of Urban agriculture in sustainable urban nutrient management". *urban agriculture, UA Magazine,. No. 23* , pp. 17-19.

215

DSCN (2002). Deuxième enquête Camerounaise auprès des ménages: première indicateurs. Yaoundé, DSCN.

DSCN (1987): Demo 1987 Vol. II, Résultat bruts, Tome 1, Rep. du Cameroun.

Dunlap, R.E. (1997). "International Opinion at the century's end: Public attitudes towards environmental issues." In L.K. Caldwell & R. Bartlett (Eds.), *Environmental Policy: Transnational issue and national trends* (pp. 201 – 225). Westport, Connecticut, London: Quorum Books.

Dunlap. E & Jones R.E. (2003). "Environmental attitudes and values." In R. Fernandez-Ballesteros (Ed), *Encyclopaedia of psychological assessment* (pp. 564 – 569).

Eleanor, L., A.R.W. Jackson and J.M. Jackson (1998). *Dictionary of Environmental Science*. Edinburgh Gate, Harlow, Essex, Longman Ltd.

Elson, D. (1995). *Male Bias in Development Process*. Manchester: Manchester University Press.

Enger. D., Smith (1998). *Environmental Science*. A Study of Interrelationships. 6[th]ed. Boston: McGraw-Hill.

Environics international and EcoLog information resources Group (2002). Environmental protection and waste management in Canada. Don Hills, Canada, Hazardous Materials Management magazine and Solid waste and recycling magazine: 59.

EPA- United States Environmental Protection Agency (2006). Municipal Solid Waste Generation, Recycling, and Disposal in the United States: Facts and Figures for 2006

EPA US (2003). "Composting: Innovative uses of compost: erosion control, tuff
remediation and landscaping." (Waste)<wysiwyg\\62\http\\www.epa.gov\epaower\non-hw\compost\> 8\6\2003.

Fombe, L.F. (2005). Substandard Housing and Slum Development in Douala: An Urban Development Perspective. Unpublished. PhD Thesis in Geography, University of Buea)

Fombe, L.F. (2006). "Solid Waste Dumping and Collection Facilities in Douala, Cameroon: Indications for a Sustainable Management of Social Sciences." Vol. xx no 1, pp. 7-18

Fombe, L.F. and Balgah, S.N. (2010). *The Urbanisation Process in Cameroon: Patterns Implications and Prospects*, Nova, New York

Fonjong, L. N. (2008). "Gender roles and practices in natural resource management in the North West Province of Cameroon." *The International Journal of Justice and Sustainability*. Special issues: communities, natural resources, and environment. Vol. 13.No. 5. Routledge

Fonjong L., Sama-Lang I. and Fombe L. (2010). An assessment of the Evolution of Land Tenure System in Cameroon and its effects on Women's Land Rights and Food Security. *Perspectives on Global Development and Technology*, Vol.9, No. 1-2., p. 154-169.

Freire, M. and R. Stren (2001). The Challenge of Urban Governance: Policies and Practices. Washington, DC: World Bank Institute

Fullerton, D. and A. Raub (2004). Economic analysis of solid waste management policies. *Addressing the economics of waste*. OECD. Paris, OECD. 1: 39-62

Furedy, C (2002). Urban Waste and Rural Farmers: Enabling Low Cost Organic Waste Reuse in Developing Countries. Recovery, Recycling and Reintegration: 6th World Congress on Integrated Resource Management.

Gender – Based analysis: A guide for policymaking. Status of women in Canada: March 1996. http://www.swc-cfc.gc.ca/publish/-e.html

Government of South Australia (GOSA) (2001). Fact sheet 1: What is the National Packaging.

Gupta, K.N. (1998). "Excreta Collection in Ghaziabad, India." In Muller, Maria. S (eds.), *The Collection of Household Excreta in Urban Low-income Settlements,* WASTE/ENSIC, Gouda/Bangkok, 1998.

Hansen, W., M. Christopher and M. Verduecheln (2002). EU Waste Policies and Challenges for Regional and local Authorities, Ecological Institute for international and European Environmental Policy. Sourced 2004.

Hardoy, J. E., D. Mitlin and D. Satterthwaite (2001). *Environmental Problems in an Urbanizing World: Finding Solutions for cities in Africa, Asia and Latin America*. London, Sterling, Earthscan publication Ltd.

Hay, M. and Sticher, S. (1984). *African Women South of the Sahara.* London: Routledge. http://www.globalissues.org/article/238/effects-of-consumerism

Hutt, D. and Swilling, M. (1999). "Johannesburg, South Africa." In Adepoju and Onibokun, A.G., ed., *Managing the Monster: Urban Waste and Governance in Africa.* Ottawa. International Development Research Centre.

Huysman, Marijk. (1994). 'Waste Picking as a Survival Strategy for Women in Indian Cities.' *Environment and Urbanisation* 6 (2).

UNEP/IETC. (1996). IETC umbrella project, project action sheet, UNEP, 15 May 1996, www.unep.org/Portals/52.Reports/k9935037IETC.doc. Accessed 17/04/2015

Imam, Mohammed, Wilson and Cheeseman (2008). "Solid waste management in Abuja, Nigeria." 28(2008), 468 – 472. www.elsevier.com/locate/wasman.

ILO/SEAPAT. International Labour Organization/South-East Asian and the Pacific Multidisciplinary Advisory Team. ILO/SEAPAT's Online Gender Learning & Information Module (accessed on 2nd February, 2012).

ISSUE (2007-2020). "Integrated support for a sustainable urban environment" www.ircwash.org/resources/integrated-support-sustainable-urban. Accessed 17-04-2015

Jane L. Parpat *et al.* (eds). (1999). *Feminism/Post modernism/ Development.* London: Routledge.

Johannessen, L. M. and G. Boyer (1999). Observation of Solid Landfills in Developing Countries: Africa, Asia and Latin America. Washington, D.C: The International Bank of Reconstruction and Development, The World Bank.

Halkier, B. (1999). "Consequences of the Politicization of Consumption: The Example of Environmentally Consumption Practices", *Journal of Environmental Policy & Planning*, Vol.1, pp. 25-41.

Lardnois and Marchand. (2000). Technical and financial performance at integrated composting waste management project sites in the Philippines, India and Nepal. Cited in Furedy, C. (2002).

Law No. 96/12 of 5th August 1996 Relating to Environmental Management in Cameroon. Ministry of the Environment and Forestry. Republic of Cameroon. "Scavenger cooperatives in Asia and Latin America." Resources, Conservation and Recycling. 31(3):229-40.

Kironde, L.J.M. (1999). "Dar es Salaam." In Adepoju and Onibokun, A.G., ed., *Managing the Monster: Urban Waste and Governance in Africa*. Ottawa. International Development Research Centre

Kabeer (1994). Reversed Realities: Gender Hierarchies in Development Thought: Verso.

Kabeer. N, Subrahmanian, R. (1996): Institutions, Relations and Outcomes: Frameworks and Tools for Gender Aware planning

Kleinman. S. (2007). "Feminist Fieldwork Analysis: Qualitative Research Methods Series 51." *A SAGE University Paper*. London. Sage Publications.

Littig, B. (2001). *Feminist Perspective on Environment and Society*. London: Pearson Education Limited.

Kumar, R. (2005). *Research Methodology: A Step by Step Guide for Beginners*. London: SAGE Publications.

Human Technology Resources (HTR) Cameroon, c. (2011). *Ministry of Urban Development and Housing (MINDU): Department of studies, Planning and Cooperation. Proposed Physical Development Plans*. Bamenda: Human Technology resources.

Luedike, Marius, K., Craig J. Thompson and Markus Giesler (2010). "Consumption as Moral Protagonism: How Myth and Ideology Animate a Brand Meditated Moral Conflict". *Journal of Consumer Research*, 36 (April)

March, Candida, Smith, I. and Mukhopadhyay, M. (1999). A Guide to Gender Analysis Frameworks, an Oxfam Publication, Chapter 1

Maclaren.V. (2003). *www.unep.or.jp/*. Retrieved from www.unep.org.

Malcom, C. (2001). Overview of EU Policy on Solid Waste Management-Implementation in EU Member States. Ukrainian National Workshop on solid waste management and disposal, Kiev, AEA Technology Environment, UK.

Manga, V. (2007). "Waste management in Cameroon: A new policy perspective?" *Resources Conservation Recycle (2007), DOI:10.1016/J.*

RESCONREC. 2007.07.003.

Markham, William T. (2008). *Environmental Organizations in Modern Germany: Hardy Survivors in the Twentieth Century and Beyond.* New York: Berghahn Books, 2008.

Markham, W T. and Fonjong, L. (2008). Environmental Organizations in Cameroon: Contributions to Environmental Protection and Civil Society. Paper presented at the International Sociological Association Forum on Sociology, Barcelona Spain, and September 5-8, 2008.

Markham, W T. and Fonjong, L. (2014). Rethinking Environmental Movements in Developing Nations: The Case of Cameroon. Joint Presentation at the ISA World Congress at Yokohama, Japan

Markham, W T. and Fonjong, L. (2012). Rethinking existing strategies for promoting women's land rights in Cameroon: Building gender capacity for male actors. Paper presented at the RC 32 session during the ISA 2nd World Forum on *Social Justice and Democratization* held at Buenos Aires, Argentina August 1-4, 2012

Markham, W T. and Fonjong, L. (2008). Funding Environmental NGOs in Cameroon.Paper presented at the RC 32 session during the ISA 2nd World Forum on *Social Justice and Democratization* held at Buenos Aires, Argentina. August 1-4, 2012

Mbuligwe S.E. & Kassenga. G.R. (2002). Potential and constrains of composting domestic solid waste in developing countries: findings from a pilot scheme in Dar-es-Salaam, Tanzania. *Resources, conservation and recycling 36 (1): 45-59.*

Medina, M. (1998). Globalization, Development and Municipal Solid Waste Management in the Third World cities. Tijuana, El Colegio de la Fontera Notre, Tijuana.

Medina, M. (2000). "Scavenger Cooperatives in Asia and Latin America." *Resources Conservation and Recycling.* 31(1): 51-69.

Medina, M. (2003). Serving the unserved: Informal refuse collection in Mexican cities. Paper presented at the CWG workshop on solid waste collection that benefits the urban poor, 8-14 March 203, Dar-es-Salaam.

Medina, M. (2008): The Informal Recycling Sector in developing countries: Organising waste pickers to enhance their impact. PPIAF (Public Infrastructure Advisory Facility). www.ppiaf.org

Mellor, M. (1997): Feminism and Ecology. New York: New York University Press.

McGuire Cathleen and McGuire Colleen (1991). Ecofeminist Visions

Micheals, A. (1966). Refuse Collection Practices. Danville, Illinois, Public Administration Service for Committee on Solid Waste American Public Works Association.

Mies M. (1993). *Ecofeminism.* London: Zed Books.

Mies, M. and Shiva, V. (1993). *Ecofeminism.* Washington: Fernwood Publications.

Mohanty *et al* (eds) (1997). Third World Women: The Politics of Feminism. Indianapolis: Indiana University Press.

Mosem, J. (1993). *Women and Development in the Third World.* London: Routledge.

Moser, O.N. (1993). *Gender Planning and Development: Theory, Practice and Training,* London: Routledge.

Moulion, J. A. (2012, 06). L'Invasion du plastique". Cameroon Tribune, pp.11.

Muller, Maria S. (1998). *The Collection of Household Excreta in Urban Low-income Settlements,* WASTE/ENSIC, Gouda/Bangkok, 1988.

Muller, M and Scheinberg, A. (2003): Gender-linked livelihoods from Modernizing the Waste Management and Recycling Sector: A Framework for Analysis and Decision Making. In *"Gender and the Waste Economy – Vietnamese and International Experiences",* V.W. Maclaren and Nguyen Thi Anh Thu (eds.), *National Political Publishing House, Hanoi,* (pp. 31 – 34).

Nana, C. (2012). *Research Methods and Applied Statistics: Beginners and Advanced Learners.* Buea: GOAHEAD.

N'dienor M. (2006). Fertilité et gestion de la fertilisation dans les systèmes maraichers périurbains des pays en développement: intérêts et limites de la valorisation agricole des déchets urbains dans ces systèmes, cas de l'agglomération d'Antananarivo (Madagascar). Thèse de doctorat, Université d'Antanarivo, Ecole Supérieure des Sciences Agronomiques, 242p+ Annexes

Neba, A. (1999). *Modern Geography of the Republic of Cameroon.* Bamenda, Camden, Neba publishers.

National Association of Professional Environmentalist. (2008). Environmental abuse in Uganda: It is lack of institutional capacity or political will? Kampala, Uganda: NAPE.

Newton, P.W. and CSIRO (2001). Australian State of Environmental Report 2001: Human Settlements Theme Report: Waste, recycling and reuse. Canberra, Department of Environmental and Heritage, Government of Australia.

Ngassa V. (2013). *Women's Land Rights Handbook, Cameroon.* London. Commonwealth Secretariat.

Njuafac. A. N (2012). The Role of Women in Solid Waste Generation and Management in Muea: Implications on the Buea council's Waste Management Strategies. BSc student project submitted to the Department of Women and Gender Studies, University of Buea.

Ngnikam E. (2000). *Evaluation environnementale et économique de systèmes de gestion des déchets solides municipaux : analyse du cas de Yaoundé Cameroun.* Thèse de Doctorat, INSA, Lyon, France.

Ngnikam. E. & Tanawa E. (2006). *Les villes d'Afrique face à leurs déchets.* Université de Technologie de Belfort-Montbeleliard. p280

Nsoh, C.C. (1994). Solid Waste Evaluation in Bamenda. Geography. Yaounde, Yaounde 1: 140.

Nsoh, F. (1998). Report on the Partners' Forum. Keep Bamenda Clean-Solid Waste Management Partnership Programme. Bamenda, COMINSUD (Common Initiative for Sustainable Development): 18

Ntantang, S. N. (2012, September and October). Plastic waste – A huge problem with possible solution. Betternews, pp.7.

OECD (2004). OECD Health Data. Paris, Organisation for Economic Co-operation and Development: 1

Olufunke Cofie, Rene van Veenhuizen, Verele de Vreede, Stan Maessen (2010, april). Waste management for Nutrient recovery: Options and challenges for urban agriculture. *The role of urban agriculture in sustainable urban nutrient management* , pp. 3-7.

222

One Hundred Worlds for Equality: A glossary of terms on equality between women and men. European Commission. Employment and social affairs. January 1998

Onibokun, A.G.; Kumuyi, A.J. (1999). "Ibadan, Nigeria." In Adepoju and Onibokun, A.G., ed., *Managing the Monster: Urban Waste and Governance in Africa*. Ottawa. Internationals Development Research Centre.

Oso, W. and Onen, D. (2008). *A General Guide to Writing Research Proposal and Report*. Kampala: Makerere University Printers.

Overholt, C.*et al.* (Eds.) (1995). *Gender roles in development projects*. Connecticut: Kumarian Press. Inc.

Pacheco, M. (1992). "Recycling in Bogota: developing a culture of urban sustainability." *Environmental and Urbanization* 4(2): 274-76.

Padilla M. (2002). Approvisionnement alimentaire des villes Méditerranéennes et Agriculture Urbaine. *In Interfaces : agricultures et villes à l'Est et au Sud de la Méditerranée*. Ed. Nasr J et Padilla M, Delta/Ifpo, 2004 : 79-94

Parrot, L., Sotamenou, J. & Dia, B.K. (2009). " Municipal solid waste management in Africa: Strategies and livelihoods in Yaounde, Cameroon". *Elsevier Ltd. Vol. 29(2009)*, 986-995.

Porter, R. C. (2004). Efficient targeting of waste policies in the production chain. *Addressing the economics of waste. OEDC*. Paris, OEDC: 116-60.

Rai, S.M. (2008). *The Gender Politics of Development*. London/New York: Zed Books.

Rumbo. J. D., "Consumer resistance in a World of Advertising Cluttr. The case of Adbulsters", Psychology and Marketing, Vol. 19(2), February 2002.

Schneider, k. (2006): *Manual for Training on gender responsive budgeting*. Eschborn: GTZ

Sarantakos. S. (1998). Social Research. London: MACMILAN PRESS LTD

Schubeler, (1996). Conceptual Framework for Municipal Solid Waste management in Low-Income countries. Vadianstrasse 42, UNDP/UNCHS (habitat)/ World Bank/SDC Collaborative Programme on MSWM in low-income countries. Published by SKAT (Swiss Centre for Development Cooperation):56.

Senkoro, H. (2003). *Solid Waste Management in Africa: A WHO/AFRO Perspective.* Solid waste collection that benefits the urban poor, Dar es Salaam, Collaborative Working Group (CWG) Workshop. World Health organization African Regional Office (WHO/AFRO).

Singh. N (1998). Think Gender in Development Research – A Review of IDRC – Funded Projects/ (1996- 1997). From a Gender Perspective, IDRC/GSD Internal Document, 1998

SKAT (2000). Planning for Sustainable and Integrated Solid Waste Management. Collaborative Working Group on Solid Waste management in Low- and Middle-income Countries, Manila, Philippines, SKAT.

Sotamenou J. (2005). Efficacité de la collecte des déchets ménagers et agriculture urbain et périurbain dans la ville de Yaoundé au Cameroun. Mémoire de DEA-PTCI en Economie, Université de Yaoundé II, Cameroun, 144p

Smith O.B., Moustier P., Mougeot L., J.A., Fall A. (2004). *Développement durable de l'agriculture urbain en Afrique francophone.* CIRAD, CRDI, 173 P. 2004

Tanawa, E., H.B. Djeuda Tchapnga, E. Ngnikam and J. Wethe (2002). La propreté dans une grande ville d'Afrique Centrale: le cas de Yaoundé au Cameroun. Collection des sciences appliquées de l'INSA de Lyon. C.B.J.-M.D.e. Henri Botta. Lyon, Presses Polytechniques et Universitaires Romandes: 123-41.

Tacoli, C. (2012). Urbanization, Gender and Urban Poverty: Paid Work and Unpaid Carework in the City. International Institute for Environment and Development: United Nations Population Fund, London, UK.

Themelis, N.J. (2002). *Integrated management of solid waste for New York.* American Society of Mechanical Engineers, New York, NAWTEC.

Thu, N.T.A (2005). "Gender and Waste in Vietnam: Integrated Waste Management in Cambodia, Laos and Vietnam", Virginia Maclaren and Tran Hieu Nhue (co-editors). *Science and Techniques Publishing House, Hanoi,* (pp. 188-198).

Tim, A. and Alan, T. (eds) (2000). *Poverty and Development into the 21ˢᵗ Century.* UK: The Open University.

Timilsina, B.P. (2001). "Public and Private Sector Involvement in Municipal Solid Waste Management: An Overview of strategy, Policy and Practices. *A Journal of the Environment* 6(7):68-77.

Tiondi, Evaline, "Women, environment and development: Sub-Saharan Africa and Latin America" (2000). *Graduate School Theses and Dissertations.* http://scholarcommons.usf.edu/etd/1549

UNPD. (2012).World Urbanization Prospects, the 2011 Revision. Department of Economic and Social Affairs, Population Division, United Nations, New York. http://esa.un.org/unup/pdf/WUP2011 Highlights.pdf Accessed: 27.08.2014

UNEP. (2000a). Geo-2000: Chapter Two: The State of Environment-Africa-Urban areas. Geo-2000: The State of the Environment, UNEP GEO-2000 Global Environment Outlook. <wysiwg://71/http://www.grida.no/geo2000/English/0059.htm>

UNEP. (2000b). Municipal Solid Waste Management. Regional overview nd information sources Africa, UNEP, Division of technology, industry and economics. http://www.unep.or.jp/ietc/ESTdir/pub/MSW/RO/Africa/topic_e.asp.2003

UNEP. (2000c). Municipal Solid Waste Management. Regional overview and information Sources Asia. Solid Waste Management Sourcebook, International Environmental Technology Centre. http://www.unep.or.jp/ietc/ESTdir/pub/MSW/ RO/content_Asia.asp.**2003**

UNEP. (2000d). Municipal Solid Waste Management. Regional overview nd information Sources North America, International Environmental Technology Centre. 2003

UNEP. (2000c). Municipal Solid Waste Management. Regional overview and information Sources. Latin America and the Caribbean. Solid Waste Management Sourcebook, International Environmental Technology Centre.

United Nations Population Division (2002). World Urbanisation Population Prospects: The 2000 Revision. Data, Tables and Highlights. New York, United Nations 2002.

UNEP. (2009a). Developing Integrated Solid Waste Management Plan: Waste Characterization and Quantification with Projections for Future. Training Manual. Vol.1. United Nations Environmental Programme. Division of Technology, Industry and Economics. International Environmental Technology Centre. Osaka/Shiga, Japan.

UNEP. (2009c). Developing Integrated Solid Waste Management Plan. ISWM Plan. Vol. 4. Compiled by United Nations Environmental Programme. Division of Technology, Industry and Economics. International Environmental Technology Centre. Osaka/Shiga, Japan.

UNEP. (2013).UNEP Global Environmental Alert Service (GEAS). Taking the pulse of the planet: connecting science with policy. www.unep.org/geas Accessed: 18.08.14).

UN-HABITAT. (2010). Collection of Municipal Solid Waste in Developing Countries. United Nations Human Settlements Programme (UN-HABITAT), Nairobi.

UNEP/UNESCO. (1995). Environmental Management in Developing Countries, Waste Management, Pub. Dresden University Vo. 2 215 pp.

UNEP. (2009c). *Developing Integate Solid Waste Management Plan:Training Manual. Volume 4: ISWM Plan.* United Nations Environment Program, 2009.

UNEP/WHO/HABITAT/WSSCC: Guidelines on municipal Wastewater Management. UNEP/GPA Coordination Office, The Hague, The Netherlands (2004)

UN-HABITAT (2008). Gender in Local Governance. A Sourcebook for Trainers. Nairobi. United Nations Human Settlements Programme

Urban Waste and Rural Farmers: Enabling Low Cost Organic Waste Reuse in Developing Countries. Recovery, Recycling and Reintegration: 6th World Congress on Integrated Resource Management.

USAID (2010). *Guide to gender integration and analysis*: http://www.transition.usaid. gov.pdf. Accessed 10-10-2012.

US EPA (2003). Summary of EPA Municipal Solid Waste Program, 2003.

UWEP (Urban Waste Expertise Programme), (1995 – 1997). Case Studies on Community Participation and Small/Micro Enterprises in Waste Management. Vers L'interdiction de certains emballages. (2012, 06). Cameroon Tribune, pp.11.

Vandava Shiva. Accessed 18, 2013.http://www.southernpress.org/authors/17

Vandava Shiva (2005). The Impoverishment of the Environment and Children Last. Environmental Philosophy: From Animal Rights the Radical Ecology (Fourth Edition). Pearson Education Inc. Upper Shadle River, NJ.

Vandava Shiva (1985). "Where has all the water gone? The case of water and feminism in India." Women and Environmental Crises. Report of the proceedings of the workshops on Women, Environment and Development, Nairobi, 10-12 July 1985). Nairobi, Environment Liaison Centre.

Visvanathan *et al.* (1997). *The Women, Gender and Development Reader.* London: Zed Books.

van de Klundert, A. & J. Ansnshutz (2000). Assessing the sustainability of alliances between stakeholders in waste management using the concept Integrated Sustainable Waste Management (ISWM). Collaborative Workshop Group on Solid waste management in Low- and Middle-income Countries, Manila, The Philippines, SKAT.

Wambo. N.I. (2013). The Role of Women in Market Waste Generation and Management: The case of Bamenda City Council Food Market, North West Region, Cameroon. BSc student project submitted to the Department of Women and Gender Studies, University of Buea.

Warren K.J. (1997). Taking empirical data seriously: An ccofeministphyilosophical perspective. In Ecofeminism: Women, Culture, Nature. Warren (ed.). Bloomington and Indianapolis. Indiana University Press

Waste/ENSIC Gorida/Bangkok, 1998.

Waugh D. (1995). *Geography: An Integrated Approach.* UK. Thomas Nelson &Sons Ltd

Webster Encyclopaedia (1977). One Volume Edition. Elsevier Copyright Management, S.A. Lausanne, Switzerland

WHO and TDR (2008).Community-directed interventions for major health problems in Africa. *A multi-country study: Final Report.* Switzerland Geneva: WHO.

WHO. (2007). Population health and waste management — Scientific data and policy options. Report of WHO workshop, Rome, Italy, 29-30 March 2007. Available at http://www.euro.who.int/__data/assets/pdf_file/0012/91101 /E91021.pdf. Accessed: 27.08.2014

World Bank. (1999). *The World Bank annual report 1999.* Washington DC; World Bank. http://documents.worldbank.org/curated/en/1999/01/437658 /world-bank-annual-report-1999

World Bank. (2012). What a Waste: A Global Review of Solid Waste Management. Urban Development Series Knowledge Papers. http://documents.worldbank.org/curated/en/2012/03/16537275/ waste -global-review-solid-waste-management. Accessed: 27/08/2014

Wilcox, D. (2001). Community Participation and empowerment: putting theory into practice. P. Notes. London, IIED: 78 – 82

Woroniuk, Berth, Helen Thomas and Johanna Schalkwyk (1997). Gender: The Concept, its Meaning and Uses – A think Piece, Department for Policy and Legal Services, SIDA, May, 1997, pg. 2

Wright, T. (2000). Report of the alternative waste management technologies and practices inquiry. Sydney, State Government of New South Wales, Office of the Minister of Environment.

YRI and Pham Bang (2003): Children and the Waste Economy in Vietnam in Cambodia, Laos and Vietnam", Virginia Maclaren and Tran HieuNhue (co-editors).Science and Techniques Publishing House, Hanoi, (pp.199-207).

Zerbock, O. (2003). Urban Solid Waste Management: Waste Reduction in Developing Nations. School of Forest Resources & Environmental Science. Michigan. Michigan Technological University: 23.

Zurbrugg, C. (1999). The challenges of solid waste disposal in developing countries. Veberlandstrasse, EAWAG/SANDEC: 28.

Appendices

Appendix I
Questionnaire

Dear respondent, the information requested below is purely for academic purpose and will be treated confidentially. The candidates is pursuing a PhD in Gender and Development Studies at the University of Buea and will very much appreciate it if you can attempt answers to all the questions where applicable. Mark an X in the appropriate box and write sentences where needed in the spaces in the spaces provided.

Section I: Household Waste Generation: Actors, activities and composition

In this section we seek to identify the different actors/partners is waste generation by gender, the household activities that generate waste and the types that households commonly generate.

1) What do you understand by waste?

2) What activities generate domestic waste in your compound/house? Who does them? State the activities in the table below, ranking them in terms of proportion of waste generated starting with the highest.

Activities	Gender of waste generator			
	Adult female	Girl child	Adult male	Boy child
Food preparation				
Sweeping of domestic environment				
Compound clearing/yard trimming				
Others (list)				
-				
-				
-				

3) State and indicate how often the waste generation activities are carried out?

Activity	Frequency of activity					
	Daily	Bi-weekly	Weekly	Bi Monthly	Monthly	Others (state)
Food preparation						
Sweeping of domestic environment						
Compound clearing / yard trimming						
Others (list)						
-						
-						

4) To what extent is the following gender (group of persons) involved in the waste generation activities carried to out: **AF-Adult female; GC-Girl Child; AM-Adult Male and BC-Boy Child**

Waste generation activities	Level of involvement of gender											
	Always				Sometimes				Never			
	AF	GC	AM	BC	AF	GC	AM	BC	AF	GC	AM	BC
Food preparation												
Sweeping of domestic environment												
Compound clearing/yard trimming												
Others (list)												
-												
-												

5) About how much waste does your household generate a day? Using "sacs and motto bag as your unit of measurement state the quantity.

Table of waste generated	Amount in "sacs and motto bags"			
	Daily	Weekly	Monthly	Other (specify)
Food preparation				
Sweeping of domestic environment				
Compound clearing/yard trimming				
Others (list)				
-				
-				

6) Do you operate a business? Yes □ (Go to a, b, c, d and e) No □(Go to 7)

a) State business activity?

_____

b) Where is the business site located? At home □ Away from home□

c) That type of product do you produce?

_____

_____

d) What type of packaging do you use?

_____

e) What types of solid waste does your business operation generate?

_____

7) Have you been prohibited (banned) advised or discouraged from the use or production of certain packaging (plastics), production or reduction of any waste types? Yes □ No □?

If yes, by who

_____

Give examples of product

_____

State reasons why.

_____

Section II: Household Waste Management practices and activities by gender

This section sets out to collect data on the different attitudes/behaviours of households in the waste management process and the involvement of men and women in waste collection, reuse and recycling activities

A. Waste Collection

1) In what type of containers do you usually collect (put, store) waste in the house? Please state

_____

2) About what size is the waste storage container your use? (Giver estimate in "sacs and motto" bags)

_____

3) Has your waste storage container got a lid/cover? Yes☐ (Go to a) No ☐ (Go to Q 4)

a) Do you use the lid regularly to cover the storage container? Yes ☐ No ☐

b) Do you seal (tie) the edge of the container

4) About how often is your house waste container emptied?

Daily ☐   Once in 2 days ☐ Once in 3 days ☐ Twice a week ☐ Once a week ☐

Others (specify) ☐

_____

5) Do you sort/separate your waste? Yes☐ (Go to a)   No☐ (Go to b)

a) If Yes i) Why do you separate your waste

_____

i) Into how many different waste types is the waste separation done? Two☐ Three☐ Three☐

List the types

_____

ii) Who does the sorting and with what level of involvement?

Waste sorter	Level of involvement			
	Always	Sometimes	Rarely	Never
Female adult				
Girl Child				
Adult Male				
Boy Child				

b)   If No, Explain why not?

_____

c)   Will you sort your waste if remunerated for or penalized for not doing so by council? Yes☐ No☐

d)   What else can motivate you to sort your waste?

_____

6)   Where do you usually deposit (throw) your waste?

_____

7)   Who transport the waste from the house to the deposal site you use and with what level of involvement?

Transporter of waste to disposal site (tick as many as applicable)	Level of involvement			
	Always	Sometimes	Rarely	Never
Adult Female				
Girl Child				
Adult Male				
Boy Child				

B.   Reuse and Recycling

Articles	Re-use	Sell	Offer	Receive as offer	
Plastics					
Glass					
Taxtile (cloth)					
Fibre (e.g bags, "sacs and mottos")					
Tins/Metals					
Others (list) -					
-					
-					

2) What do you do with the food wastes, leaves and yard trimmings that come from your domestic environment?

_____

_____

Section III: Council Waste Management Strategies And Gender Gaps Therein

In this section, information is sought to evaluate the effectiveness of management policies/strategies and identify the gender gaps that such strategies create or do not address.

Waste storage and collection

1) How close is the waste collection site to your house? About one electric pole☐ About two poles☐ more three poles☐ About four poles☐ About five poles☐ More than five poles☐

2) What kind of waste collection site is it? Mobile van☐ Waste bin☐ Open space☐

Others (specify)

_____

3) How often is the public waste collection site closet to your house empties by the council or any other formal (NGO, private company) waste agent?

1. Daily☐ 2. Once a week☐ 3. Twice a week☐ 4. Thrice a waste☐ 5. Never ☐

others☐ (specify)

_____

4) What is usually the state of the waste collection site/transfer/public bin?

1. Empty ☐ 2. Good ☐   3. Always full ☐   4. Always overfull☐

5) How would you best evaluate the collection and transfer process of waste at the waste collection/transfer station/public bin?

1. Very good ☐   2. Good ☐   3. Fair ☐ 4. Average ☐   5. Bad ☐

6) Are you already to pay for waste to be collected from your house according to your schedule? Yes ☐   No ☐.   If Yes ☐ (Go to a) No (Go to b)

a) If yes, how much are you ready to pay a month for waste collection form your house at your convenience? State amount in FCFA _____

b) If No, Explain why?

_____

7) How would you describe the attitude of the waste collection team/ personnel to the public?

_____

8) Who decided on the placement (location) of the public bin/pen dump that you use?

1. City council ☐ 2. Quarter people ☐ 3. Quarter people in collaboration with council☐ 4. Private operator in collaboration with council can people ☐ 5. Private operator ☐ 6. Do not know☐

9) How would you best evaluate the state and convenience of the collection site/public bin/mobile van?

-	1.	Adequate in size	Yes □	No □
-	2.	Adequate in height for all	Yes □	No □
-	3.	In a good state	Yes □	No □
-	4.	Well placed by the street	Yes □	No □
-	5.	Accessible	Yes □	No □

10) How do you evaluate the state of solid waste management in the city over the six months?

1. Has improved □ (Go to Q. 10). Remains the same □ (Go to 12). Has deteriorated □ (Go to Q. 11)

11) If the situation has improved.

a)   State how?

_____

b)   What in your opinion has contributed to this improved?

_____

12) If the situation of the waste management service has deteriorated,

a)   How

_____

b)   What reasons can you suggest for the deteriorated state?

_____

13) How do you think the following partners can contribute in improving the waste management situation?

a)   Waste generators (households)

_____

_____

b)   The council

_____

_____

c)   Private waste operators

_____

_____

d)   Others (specify)

_____

_____

Technology

14) How adaptable are the waste bins to the users in terms of height, size and shape?

Structure	Very adaptable	Adaptable	Uncertain	Not very	Not at all
Height					
Size					
Shape					

15) How much is the distance between your house and the public waste bin, if that is the service you use?

N.B. The distance between one pole and the other is 50m

1. Distance of about one electricity pole ☐
2. Distance of about two electricity poles ☐
3. Distance of about three electricity poles ☐
4. Distance of about four electricity poles ☐
5. Distance of about five electricity poles ☐
6. Distance of more than five electricity poles ☐

16) By what method is the waste container usually transferred from the house to the place of deposition?

Place of waste Deposition	Technology used				
Council Public Van	Head/Hand	Wheel Barrow	Push-push (Truck)	Motorized vehicle	Others
Council Mobile Van					
In the compound					
Open space out of the compound					
River (stream side)					
Road side					
Gutter					
Others (specify)					

17) Why have you chosen to deposit your waste where you do? State a maximum of two places you use the most in order of highest frequency to the least, and state the reason why?

Place 1: _____

Reason: _____

Place 2: _____

Reason: _____

18) Do you carry out any composting? Yes ☐   No ☐. If Yes ☐ (Go to 17) No ☐ (Go to 18)

19) If Yes why?

_____

20) If No, why not?

_____

Waste Management and Environmental Concerns

1) Have you ever noticed the following in and around the waste storage container at your home and how often?

Tick the appropriate response

Characteristics	Frequency of occurrence			
	Always	Sometimes	Rarely	Never
Dark flowing water				
Odour				
Mosquitoes/ flies				
Cockroaches				
Fire				
Domestic Animals				
Rats				
Scavengers				
Flies				
Others (Specify)				

2) Have you ever noticed the following in and around the waste storage container at the public waste collection/ transfer station? Tick the appropriate response.

Characteristics	Frequency of occurrence			
	Always	Sometimes	Rarely	Never
Dark flowing water				
Odour				
Mosquitoes/ flies				
Cockroaches				
Fire				
Domestic Animals				
Rats				
Scavengers				
Others (specify)				

3)   (a) Have you witnessed the presence of waste dumped in rivers, streams, valleys and gutters or road sides?

Yes ☐   (Go to Q. 2.b)   No ☐

(b) Are the nearby streams/ rivers where waste is dumped used by the population around or downstream?

Yes ☐   (Go to Q. 2.c)   No ☐

(c) If yes, for what purposes? Drinking ☐   Washing ☐

Cooking☐   Gardening ☐

Others ☐ specify _____

(d) Who uses this water resource more? Men ☐   Women ☐

4)   Have you experienced the following in the past one year?

(a)   Used insecticides: Yes ☐   No ☐

(b)   Used raticides: Yes ☐   No ☐

(c)   A flood over the past year which you relate to waste effect on drainage paths: Yes ☐   No ☐

(d)   Common illnesses which you can attribute to poor waste disposal?

Yes ☐   No ☐

Others (specify) _____

5)   How appropriate do you think is the location of an open waste dump site, transfer station/ collection site in terms of distance from home settlements and activities?

Distance from	Degree of appropriateness			Justification for judgement (Why do you think so?
	App	Inapp	Uncertain	
Homes				
Cooked Food eating places				
Fresh meat sale stands				
Instant edible fruit sales stands				

App – Appropriate   Inapp – Inappropriate

6) Have you ever had any sensitization/ education on waste management?
Yes ☐   (Go to a, b, c and d) No ☐ (Go to section IV)
(a) How?

_____

(b) By who?

_____

(c) What was the main message?

_____

Section IV: Mainstreaming Gender In Municipal Waste Management Policies/ Strategies

In this section, we solicit information on gender issues that can serve as a guide in mainstreaming gender in municipal waste management policies/ strategies

1) Does the distance from your residence to the closest council authorized bin influence your effective/ convenient use of it? Yes ☐ (Go to a) No ☐
a) How?

_____

b) To what extent? Very strongly ☐   Strongly ☐   Not certain ☐
Not strongly ☐ Not at all ☐
2) Do you or have you had any opportunity of offering your opinion as to what waste management strategy the council should offer? Yes ☐No ☐
3) Given the opportunity, what would you propose as a better measure/ method that can best serve your situation in relation to council waste collection

techniques, equipment, frequency of service delivery etc.?

_____

_____

4) Do you pay for waste collection services? Yes (Go to Q. 5)   No ☐ (Go to Q. 6)

5) If yes, how often do you pay and how much? Tick the frequency and state the amount.

Frequency of payment		Amount
Daily		
Weekly		
Monthly		

Appendix II

Household On-Site Waste Quantity Measurement Information Sheet

Subdivision: Bamenda I ☐   Bamenda II ☐

Bamenda III ☐

Quarter/Neighbourhood_____

Household

Size_____

Date_____Day I ☐   Day II

☐

Waste component	Yes	No	Waste quantities	Waste Volumes
Food waste				
Paper/cardboard				
Plastic				
Textile				
Rubber				
Leather				

Yard waste (compound clearing)				
Glass				
Metals				
Fibre				
Others:				

Observations/Remarks---

Appendix III
Market Information Sheet

Market Items/Services Coverage Census Sheet
Name of Market_____

Food Items
➢ Vegetables _____
➢ Tubers_____
➢ Cereals/Grains_____
➢ Tree crops_____
➢ Fruits_____
➢ Canned/packaged food_____
➢ Others_____

Non-Food Items
➢ Wears_____
➢ Household utensils_____
➢ Automobile Parts_____
➢ Cosmetics_____
➢ Non-food groceries_____
➢ Textile_____
➢ Electronic appliances_____
➢ Others(Specify)_____

Activities/Services
➢ Sewing/Embroidery_____

- ➢ Mending/Repairs_____
- ➢ Milling_____
- ➢ Restoration/Eating places _____
- ➢ Hair dressing_____
- ➢

rinking Places(Bars)_____

- ➢ Others (Specify)_____

Packaging material
- ➢ Plastics_____
- ➢ Carton/paper_____
- ➢ Leaves_____
- ➢ Glass_____
- ➢ Metal_____
- ➢ No packaging_____
- ➢ Others (Specify)_____

Appendix IV

Structured Interview Guide for Market Vendors

Market_____

1. Age: Less than 20 ☐ 20-29 ☐ 30-39 ☐ 40-49 ☐ 50 and above ☐
2. Sex: Female ☐ Male ☐
3. Marital status: Married ☐ Single ☐ Widow (er) ☐
4. Level of Education: Non ☐ Primary ☐ Secondary ☐ Post Secondary ☐
5. Items sold/Activities: Food Items ☐ Non-Food Items ☐

Activities/Services ☐

Food Products
- ➢ Vegetables
- ➢ Tubers
- ➢ Cereals/Grains
- ➢ Tree crops
- ➢ Fruits
- ➢ Canned/packaged food
- ➢ Others

242

Non-Food Items
➢ Wears
➢ Household utensils Automobile Parts
➢ Cosmetics
➢ Non-food groceries
➢ Textile
➢ Electronic appliances
➢ Others(Specify)

_____

Activities/Services
➢ Sewing/Embroidery
➢ Mending/Repairs
➢ Milling
➢ Restoration/Eating places
➢ Hair dressing
➢ Drinking Places(Bars)
➢ Others

_____

Packaging: Plastic☐ Carton/paper ☐ Leaves ☐ Glass ☐ Metal ☐No packaging ☐

6.
Perception of waste:_____

7. State other activities you do that generate waste while you are in the market:

_____

8. In what type of container do you store your waste before disposing of it?
_____

9. For how long do you store the waste?_____Why that length of time?_____

10. How much waste do you generate daily ("sacs and motors" bag measurement)_____

11. What do you do with the waste you generate? _____

_____

12. Do you have difficulties managing your waste? No ☐ Yes ☐

Why?_____

13. Attempted solutions to problems:_____

14

Market waste collection: By who?_____

How often?_____

15.

Suggestions for improvement:_____ _

_____

16 Have you been sensitized on waste Management in the market? Yes ☐ No

☐

If yes, by who?_____

What was the message:_____

17. Are you willing to separate your waste? Yes ☐ No Why?

_____

Observation

- Number of council bins in and around:--

- Attitude/behaviour towards interview:--

- Others --

Appendix V

Interview Guide for Council Authorities And Workers

1   What do you understand by waste?

2   Which are the main sources of municipal solid waste in Bamenda?

3   How much waste is produced by the municipality?

4   Can you give an estimate of the waste quantities and waste types by source of waste?

5   What is the proportion of household and market waste to the overall in terms of types of waste and quantities produced?

A.   *Assessing Council Policies on Municipal Solid Waste Management*

- Laws/Acts (International or national laws, Environmental protection Law or Clean Act)
- Regulations/standards (on the construction and operation of landfills, incinerators and composting plants)
- Economic instruments (Financial disincentives such as charges, levy, penalties and fines for waste generators or economic incentives such as subsidies or pay back for recycling)
- Enforcement mechanisms (Strategies)

To what extent do the above elements guide the Bamenda City Council waste management policies with regard to:

1   Waste reduction at source

2   Waste separation at source

3   Primary storage and collection of waste

4   Transportation and transfer stations

5   Treatment of waste

6   Landfills

7   Recycling

8   Resource recovery

9   Different types of waste identified above?

B.   *Assessing waste management related institutions*

1   Do you know of other waste management operators functioning within the Bamenda City Council Area?

2   If yes, name them and state in what activities of waste management they are involved.

3   How do you relate with each of these operators?

4   How would you evaluate the performance of each operator?

C. Assessing financing mechanisms

Which mechanisms for financing waste management are available to the council?

D. Assessing technology for environmental sound practice and convenience to users

Type of waste service	Technology		
	Type	Number	Important Features
Collection	E.g. Dust bins	5	Partition for different waste types
Transportation			
Treatment			
Disposal			
Recycling			

1. Do you have public waste bins situated in the city?
2. If yes, what determines the location sites?
3. Are bins labelled to indicate waste separation? Is there any provision for waste separation at the bins?
4. How far apart is one dust bin away from the other? Which waste disposal technique is used by the council?
5. How far is the disposal site away from settlements?
6. What problems do you encounter with the location?
7. How many of them are they?
8. What are the possible adverse effects of the Dump site/landfill standard on the immediate environment and the population?
9. How would you evaluate the technical and quality standards of the disposal waste system in use?
10. Are there prospects for improvement?
11. How well does the current waste management system meet environmental concerns?
12. What predictions about waste quantities and management are there for the future?

13. What prospects are there for abandoned waste sites—scavenging and waste recovery?

E. *Assessing Stakeholder Participation*

1. Who are the major stakeholders in MSWM?

2. What is the level of stakeholders' participation in SWM decision making such as setting the level and type of MSWM services (e.g. door to door collection or location of transfer station?

3. What measures do you use to improve stakeholders participation (e.g. materials, campaigns, meetings, political and social interaction)?

Gender issues in municipal solid waste management

1 How is gender taken into account in waste management politics and strategies regarding

a) Participation of households (men and women) in decision making with regard to the location of public bins and the location of waste disposal site (dump site).

b) The participation of market vendors (men and women) in decision making with regard to the location of public bins and the location of waste disposal site (dump site).

c) At what level of decision making are men and women in waste management and with what representation?

d) Recruitment of waste workers for collection, transportation, landfilling etc

e) Participation of waste workers (men and women) in waste decision making issues

f) Gender access to scavenging waste resources for livelihood

g) The impact of waste management techniques on the health of families?

Recommendations for mainstreaming gender in MSW management

1 Do waste generators attitudes and their waste management practices (behaviour) affect the efficiency of council waste policy/strategies? If so, how is the council using this knowledge to improve on its waste management strategies?

2 Do you see women as main actors in household and market waste generation and management? If yes, how has the council used or intend to use this information as an opportunity to promote efficiency in MSWM. (N. B. Verify response to IV-1 above).

Appendix VI

Some Code Grounding Quotation Reports For Questions On Household Waste Survey Questionnaire

Reason why banned or discouraged from the use or production of certain packaging, production or reduction of any waste

Table: Reason why banned or discouraged from the use or production of certain packaging, production or reduction of any waste

Code	Code description	Grounding	Quotation
Food contamination	Plastics or some metal containers generate health problems as cancers when used to package food because they contain harmful chemical substances. Water can also be contaminated	58 (57.4%)	"Plastics not good. Contain substance or chemical which is not good for our health like those used to tie fufu corn" [Female taxation officer, post secondary level, aged between 36-40] "Some chemical substances called dioxin may get into water under pressure from heat. This substance is detrimental to our health" [Male civil servant, post secondary, aged between 41-45] "Carcinogen, that is cancer causing substances" [Male doctor, post secondary, aged 36-40]
Non-degradable	Plastic papers and containers resist natural degradation	26 (25.7%)	"Plastics papers and containers deteriorate our soil since it does not decompose" [Female, business and

248

	processes thus polluting the soil		teaching, post secondary level, aged 31-35] "It prevents the easy accessibility of crop roots to the soil, it also makes it difficult for the micro-organism of the soil to work well" [Female business woman, postsecondary, aged 21-25]
Pollution after burning	Plastics generate harmful pollutants when burned	6 (5.9%)	"That burning plastics as method of disposal produces odors that pollute the atmosphere and are potentially dangerous to human health" [Female teacher, post secondary, 26-30]
Water pollution	Harmful to aquatic life	2 (2.0%)	"Not good to be thrown in the streams as it is harmful to aquatic life" [Male secondary education, unemployed, aged 26-30]
Dirty/Loss of aesthetics	Dirty the environment	7 (6.9%)	"It is because where it is been used and thrown around, it dirties the environment" [Female, reverent sister, post secondary, aged 16-20]
Jealousy	Ban out of jealousy	1 (1.0%)	"Mere jealousy" [Female, post secondary education, aged 16-20]
I don't know	I don't know	1 (1.0%)	

N (total responses) =111.

Reason why waste is separated

Code	Code description	Grounding	Quotation
Manure	Organic waste are separated to be used as manure	42.1%	"It goes mostly to the farm" [Male, retired worker, aged above 50]
Degradable ability	To separate degradable waste from non-degradable ones as to make sure that non degradable ones are specially disposed	23.0%	"To burn plastics and throw which can decay in the farm" [Female teacher, post secondary, age above 50] "In order to burn plastics" [Male driver, primary level, aged above 50] "To separate biodegradable from non-biodegradable" [Female teacher] "Some easily get rotten" [Female, secondary, aged 31-35]
Nature of waste	To separate dust from bigger particles, or to avoid smell that may be generated by some waste	9.5%	"Because some waste may be in dust form" [Female, business woman, post secondary, aged 16-20] "Some is water and may soak others" [Female business woman, post secondary, aged 21-25] "Some of the waste undergo decay fast" [Male, post secondary, aged 36-40] "Some toilet waste don't fit being kept together" [Male farmer, aged 26-30]
Injury	To avoid waste that can harm or attract flies, bad odour or contaminants	4.0%	"To avoid waste that can harm someone" [Female student, postsecondary, aged 26-30]

			"To keep away flies and mosquitoes" [Adult Male business man, post secondary] "Because of broken glasses" [Adult male journalist, primary education] "Because if some are mixed it will be poisonous" [Male driver, secondary, aged 31-35] "Avoid odour" [Male farmer, primary, aged 26-30]
Usage/recycling	Some will be used for other purposes	7.1%	"Some at times are still useful" [Female traditional Doctor, primary education, aged 41-45] "Because I use the plantain peals for niki (local potash)" [Female, farmer and business woman, primary, aged above 50]
Feed for animals	Use to feed animals like pigs	4.0%	"Some are used as food for pigs" [Female, post secondary, trader, aged 21-25]
Burning	To burn flammable waste like paper	10.3%	"To burn paper" [House wife, secondary, aged 36-40] "So as to burn paper that do not decay" [Female business woman, primary, aged above 50] "Some waste are burned while others are thrown in the farm" [Adult female trader, post secondary]

N (number of responses)=126

Reason why waste is not separated

Code	Code description	Grounding	Quotation
No need	Importance of sorting not perceived, waste considered useless, sorting considered as a waste of time	42.1%	"I see no need" [Female, post secondary, aged between 31-35] "No need sorting since we dump everything" [Female, taxation officer, post secondary, age between 36-40] "Because I consider waste as waste, there is no need for separation" [Male, office worker, post secondary, aged 31-35] "It is a mere waste of time" [Male civil servant, aged 36-40] "It is waste of time" [Male, no occupation, primary level, aged 16-20]
Time constraint	Lack of time, time consuming	23.0%	
Lack of labour	Lack of labour force i.e. no one to do it	9.5%	
Not informed	Lack of awareness	4.0%	
Council responsibility	Believe it is the responsibility of the council	7.1%	
Not required by collectors	Because collectors do not consider sorting	4.0%	"There is no need because the collectors do not consider sorting" [Female teacher, post secondary, aged 46-50]
No reason		10.3%	

Not a resource/Waste	Waste is waste, so no need sorting, not considered as a resource	42.1%	"Waste is waste, so we cannot separate them" [Male builder, primary, aged 16-20] "Because anything rejected is waste" [Female decorator, Primary, aged above 50]
No farm	Absence of farm land	23.0%	
No space		9.5%	
No machine	Lack of machine to sort waste	4.0%	"Because there is no machine to sort waste" [Male, civil servant, post secondary education, aged 31-35]
Burdensome	Burdensome, laziness	7.1%	
Not convenient	Because of odours, it is not convenient to sort	4.0%	"It has odour, hence difficult to sort" [Female homemaker, secondary, aged 31-35]
Containers	Lack of sufficient storage facilities	10.3%	"Because I don't have another storage container" [Male nurse, post secondary, aged 36-40]
Lack of interest		42.1%	
Uniformity	No uniformity in term of usefulness or uselessness	23.0%	"All is manure" [Female teacher, post secondary, aged 41-55] "Because all of them are waste" [Female, aged 26-30]

N (number of responses)=126

What else can motivate you to sort your waste?

253

Code	Code description	Grounding	Quotation
Sensitization	Sensitization on the importance and use of sorting, and on how to sort	24.1%	"If the council encourages sorting" [Female teacher, post secondary, 46-50] "A clear explanation" [Male painter, secondary education, aged 26-30]
Health	To curb health hazard related to improper waste disposal as malaria	26.5%	"Health hazard of waste" [Male doctor, post secondary, 36-40] "Disease like malaria from mosquitoes" [Male driver, secondary, aged 31-35] "Avoid some diseases and flies" [Female housewife, secondary education, aged 26-30]
Council request	If containers are provided for and if council collects waste separately	.6%	"Storage containers" [Male driver, post secondary, aged 21-25] "If council collects them separately" [Male consultant, postsecondary, aged 41-45]
If carried daily	If waste is carried daily because of storage constrain	1.8%	"If they carry it everyday" [Adult male journalist, primary]
Pit usage	To prolong the age of the pit by depositing only decomposable waste.	5.3%	"So as not to fill the pit too quickly' [Female, Post secondary, aged between 41-45]
Time availability/	Time and labour	2.9%	"If I have a helping

labour	availability		hand" [Female public service worker, post secondary, aged 36-40]
Economic value	If waste is attributed economic value. Because they can be sold or if they are deemed useful	7.6%	"If some of the items are purchasable" [Female, post secondary, aged between 26-30] "If it has use to one" [Female teacher, aged above 50] "To sell emptied containers" [Female aged 21-25]
Gift	To offer as gift to someone who uses it e.g.; waste food given to someone who rare pigs.	19.4%	"To give as offer" [Female housewife, primary, aged 41-45]
Bins	If there are bins provided for the various types of waste	2.4%	"Placing bins around" [Male driver, primary level, aged above 50]
Hygiene and environmental protection	Love for hygiene and for environmental protection as some waste are not biodegradable, generate odours, mosquitoes, flies	1.2%	"Fear of mosquitoes" [Female, business woman, post secondary education, aged 16-20] "Avoid odour" [Female teacher, post secondary, aged 26-30]
Ease disposal	This eases waste disposal and reduce pollution risk	.6%	"Eliminate easily to avoid pollution" [Retired female, post secondary aged between 41-45]
Recycling	The existence of	2.4%	"Recycling industries if

	recycling industries		"they are set up" [Male civil servant, post secondary education, aged 41-45]
Manure/farm	If it can be used as manure or if a farm is available for the application of manure	.6%	"If I have a farm to deposit compost waste" [Female teacher, post secondary, aged 36-40] "If it is to be used as manure" [Female teacher, secondary, aged 21-25]
Degradable ability	To separate degradable waste from non-degradable ones as to make sure that non degradable ones are specially disposed	2.9%	"Some do not decay" [Female retired civil servant, post secondary, aged above 50]
Emotional contagion/popular action/ad populum	If other people practice waste separation	.6%	"If other people do the same" [Female call box, post secondary, aged 41-45]
Nothing/Already a tradition	Nothing, already inculcated as habit because one is conscious of the importance	1.2%	"Sorting is already a practice in my home" "Nothing because it is necessary' [Male electrician, postsecondary, aged 21-25]
Quantity of waste	If the waste amounts are small	24.1%	
If paid	Financial incentives/reward	26.5%	"Only payment can motivate me" [Male teacher, post

			secondary, aged 25-26]
Burn	To burn waste that are easy to burn	.6%	"Easy to burn" [Female police officer, aged 21-25]
Penalty/exigency	If punished or compelled to do so	1.8%	
Ease disposal	Ease disposal	5.3%	"Easy to empty when time comes" [Female trader, post secondary, aged 21-25]
Animal husbandry	If animal around as to consume waste food	2.9%	"If I have pigs I will separate the waste" [Female, post secondary, aged 26-30]
I don't know	I don't know	7.6%	
Injury	To remove potentially harmful material like metals or bottles etc.	19.4%	"To avoid injuries" [Male business man, post secondary, aged 26-30]
Nature of waste	To separate dust from bigger particles, or to avoid smell that may be generated by some waste	2.4%	"If it contains plastics" [Female decorator, primary, aged above 50]
Will power	Will power	1.2%	

N (number of responses)=197

What in your opinion has contributed to the improvement?

Perceived factors that have contributed to improvement

Code	Code description	Grounding	Quotations
Don't know		1.8	
Public collaboration	People now use public waste facilities made available. There is improved awareness on environmental hygiene	9.8	"The growing awareness on the need to have a clean environment" [Male, retired, above 50 years] "The citizens are sensitized of pollution so they help the council to gather them" [Female retired worker, post secondary education, aged between 41-45] "By using the bins correctly when I am opportune to" [Female, aged 21-25] "Collaboration b of both council and quarter people" [Female administrator, aged 41-45] "Complains received from people and the council has adjusted" [male trader, secondary, aged 36-40]
Improved services	Improvementin waste collection and disposal facilities and mechanism (efficiency of service and more people employed) by the city council	52.7	"Employment of cleaning staff" [Female, post secondary, taxation officer, aged 36-40] "Availability of dust bins" [Female, post secondary education, teacher, aged between 31-35] "Availability of council waste collection vans, waste deposit site and personnel to clear out waste into the van" [Male, office worker, post secondary, aged 31-35] "The council workers are doing their best to clean the town" [Female, retired teacher, post secondary, aged above 50]

258

			"The presence of HYSACAM" [Female teacher, post secondary, aged 26-30] "The mayor and the government delegate in the city council" [Female, student teacher, post secondary, aged 21-25] "People sweep the street" [Female trader]
Political ambitions	Political ambitions	.4	"The council is trying to keep their promise so that can be voted for the upcoming election" [Male teacher, post secondary, aged 21-45]
Adequate disposal	Waste is thrown far away from residential areas	.9	"Because it is thrown in very far places" [Female nurse, aged 26-30] "A need dump site has been acquired" [Female, secondary, aged 31-35]
Clean up campaign	Monthly cleanup campaign termed "Operation keep Bamenda clean"	2.7	
Holiday jobs	Holiday jobs are provided to the youth to clean up the town	.4	"Holiday jobs for youth to collect the waste" [Female, aged 21-25]
Change of leadership	Change of authorities	1.8	"Change of municipal authority" [Male civil servant, post secondary, aged between 41-45] "The new mayor has the town at heart and wants to improve things" [Female nurse, post secondary,

			aged 26-30]
Efficiency of authority	Efficiency of authority and good planning	7.6	"Council and government delegate" [Female farmer, no education, 46-50] "The city council" [Female, trader, post secondary, aged 21-25] "Complains received from people and the council has adjusted" [male trader, secondary, aged 36-40] "The delegate at the Bamenda city council" [Female decorator, primary, aged above 50] "Channel of complain reach the rich decision makes" [Male teacher, postsecondary, aged 46-50] "Planning" [Female nurse, post secondary, aged 26-30]
Administrative collaboration	Collaboration between local authorities and the council	.4	"Collaboration between council and quarter head" [Male tailor, post secondary education, aged 41-45]
Workers' motivation	Workers are well motivated	1.8	"The council workers are doing their best and the council too is motivating them" [Male pensioner, post secondary, aged above 50]
Workers' devotion	The devotion and good work done by the workers	6.3	"Commitment of the council's workers" [Female Sonel Worker, secondary, aged 21-25]
Availability of finance/loan	Council collect taxes which makes finance available, contribution from people,	4.0	"As a result of revenue collected by the council" [Business woman, post secondary, aged 26-30] "I have contributed money to the congress development" [Female

	increase in fund, provision of loan		business woman, postsecondary, aged 21-25] "Loan to the council" [Male teacher, post secondary, aged 31-35] "Increase in population and income to the council" [Female farmer, postsecondary, aged 31-35]
Sensitization	Public sensitization on environmental issues	2.2	"Sensitization and accessibility to mobile van public bins" [Female student, post secondary, 26-30]
Consciousness	Public awareness and consciousness	.9	"Diseases have made people to know the important of keeping the environment clean" [Male civil servant, post secondary, 36-40] "The awareness of everybody about keeping the environment clean" [Male businessman, post secondary, aged 26-30]
Public demand	Public request for better waste management	.9	"Public pleas to the council" [Female teacher, above 50] "Channel of complain reach the rich decision makes" [Male teacher, postsecondary, aged 46-50]
Penalty	Penalty on defaulters, council police to track defaulter	.9	"The sanctions imposed by council on defaulters" [Female teacher, post secondary, 26-30] "Council police" [Male business, secondary aged 36-40]
Private collaboration	Collaboration from NGOs and other private operators	.9	"NGOs" [Male builder, primary, aged 16-20] "The city council, private operators in collaboration with the council" [Male driver, post secondary, aged 21-25] "The employment of the services

			of four contractors to clean the town" [Female contract officer, post secondary, aged 46-50]
Problems	Council awareness of the effects of poor waste disposal on the environment and public health: flood, health problems (malaria, cholera), waste accumulation and amount due to increase population	3.6	"The disaster that waste has being causing in gutters, which caused flood" [Male unemployed, secondary, aged 26-30] "Riots from the population and dumping of waste on road sites" [Female trader, secondary, aged 21-25] "The dumping of waste on streets and rivers" [Female housewife, post secondary, aged 31-35] "I think it is to reduce the level of malaria in the society" [Male nurse, post secondary, aged 36-40] "There are more households now than before" [Female vendor, aged 21-25]

N (number of responses) =224

Perceived factors that have contributed to improvement: distribution by sexFactors that have contributed to improve		What sex do you belong to		Total
		Male	Female	
Don't know	n	3	1	4
	%	3.8%	.9%	2.1%
People now use public waste facilities made available. There is improved awareness on environmental hygiene	n	5	16	21
	%	6.3%	14.8%	11.2%
Improvement in waste collection and disposal facilities and mechanism (efficiency of service and more people employed) by the city council	n	46	67	113
	%	58.2%	62.0%	60.4%
Political ambitions	n	0	1	1
	%	.0%	.9%	.5%
Waste is thrown far away from residential areas	n	1	1	2
	%	1.3%	.9%	1.1%
Monthly cleanup campaign termed "Operation keep Bamenda clean	n	6	0	6
	%	7.6%	.0%	3.2%
Holiday jobs are provided to the youth to clean up the town	n	0	1	1
	%	.0%	.9%	.5%
Change of authorities	n	3	1	4
	%	3.8%	.9%	2.1%
Efficiency of authority and good planning	n	8	7	15
	%	10.1%	6.5%	8.0%
Collaboration between local authorities and the council	n	1	0	1
	%	1.3%	.0%	.5%
Workers are well motivated	n	3	1	4
	%	3.8%	.9%	2.1%
The devotion and good work done by the workers	n	4	10	14
	%	5.1%	9.3%	7.5%
	n	3	6	9

Council collect taxes which makes finance available, contribution from people, increase in fund, provision of loan	%	3.8%	5.6%	4.8%
Public sensitization on environmental issues	n	1	4	5
	%	1.3%	3.7%	2.7%
Public awareness and consciousness	n	1	1	2
	%	1.3%	.9%	1.1%
Public request for better waste management	n	1	1	2
	%	1.3%	.9%	1.1%
Penalty on defaulters, council police to track defaulter	n	0	1	1
	%	.0%	.9%	.5%
Collaboration from NGOs and other private operators	n	1	1	2
	%	1.3%	.9%	1.1%
Council awareness of the effects of poor waste disposal on the environment and public health: flood, health problems (malaria, cholera), waste accumulation and amount due to increase population	n	6	2	8
	%	7.6%	1.9%	4.3%
Total	n	79	108	187
	%	42.2%	57.8%	100.0%

Perceived factors that have contributed to improvement: distribution by subdivision

Factors that have contributed to improve		Subdivision			
		Bamenda I	Bamenda II	Bamenda III	Total
Don't know	n	0	4	0	4
	%	.0%	3.4%	.0%	2.0%
People now use public waste facilities made available. There is improved awareness on environmental hygiene	n	2	16	4	22
	%	22.2%	13.8%	5.6%	11.2%
Improvement in waste collection and disposal facilities and mechanism (efficiency of service and more people employed) by the city council	n	4	64	50	118
	%	44.4%	55.2%	70.4%	60.2%
Political ambitions	n	0	1	0	1
	%	.0%	.9%	.0%	.5%
Waste is thrown far away from residential areas	n	0	1	1	2
	%	.0%	.9%	1.4%	1.0%
Monthly clean up campaign termed "Operation keep Bamenda clean	n	1	2	3	6
	%	11.1%	1.7%	4.2%	3.1%
Holiday jobs are provided to the youth to clean up the town	n	0	1	0	1
	%	.0%	.9%	.0%	.5%
Change of authorities	n	0	4	0	4
	%	.0%	3.4%	.0%	2.0%
Efficiency of authority and good planning	n	2	13	2	17
	%	22.2%	11.2%	2.8%	8.7%
Collaboration between local authorities and the council	n	0	0	1	1
	%	.0%	.0%	1.4%	.5%
Workers are well motivated	n	0	2	2	4
	%	.0%	1.7%	2.8%	2.0%
The devotion and good work done by the workers	n	0	2	12	14
	%	.0%	1.7%	16.9%	7.1%
	n	0	6	3	9

Council collect taxes which makes finance available, contribution from people, increase in fund, provision of loan	%	.0%	5.2%	4.2%	4.6%
Public sensitization on environmental issues	n	0	2	3	5
	%	.0%	1.7%	4.2%	2.6%
Public awareness and consciousness	n	0	1	1	2
	%	.0%	.9%	1.4%	1.0%
Public request for better waste management	n	0	1	1	2
	%	.0%	.9%	1.4%	1.0%
Penalty on defaulters, council police to track defaulter	n	0	0	2	2
	%	.0%	.0%	2.8%	1.0%
Collaboration from NGOs and other private operators	n	0	2	0	2
	%	.0%	1.7%	.0%	1.0%
Council awareness of the effects of poor waste disposal on the environment and public health: flood, health problems (malaria, cholera), waste accumulation and amount due to increase population	n	0	7	1	8
	%	.0%	6.0%	1.4%	4.1%
Total	n	9	116	71	196
	%	4.6%	59.2%	36.2%	100.0%

Printed in the United States
By Bookmasters